AF344418

CORROSION AND ITS PREVENTION IN WATERS

Corrosion and its Prevention in Waters

G. BUTLER, M.A., Ph.D., and
H. C. K. ISON, A.I.M.

ROBERT E. KRIEGER PUBLISHING COMPANY
HUNTINGTON, NEW YORK
1978

Original Edition 1966
Reprint 1978

Printed and Published by
ROBERT E. KRIEGER PUBLISHING COMPANY, INC.
645 NEW YORK AVENUE
HUNTINGTON, NEW YORK 11743

Library of Congress Cataloging in Publication Data

Butler, George, 1924-
   Corrosion and its prevention in waters.

Reprint of the 1966 ed. published by Reinhold Pub. Corp., New York
   Bibliography: p.
   Includes index.
   1. Corrosion and anti-corrosives. I. Ison, Hugh Colin Kinder,
   joint author. II. Title.
   TA462.B88   1977   620.1'1223   76-30515
ISBN 0-88275-515-3

Printed in the United States of America.

*To Joan and Margaret who, throughout such a
long period of neglect, have helped so much by
their patience and understanding.*

Preface

Over a period of many years engaged in the experimental investigation of corrosion phenomena in the laboratory and in the diagnosis of industrial corrosion problems we have become more and more aware that a large majority of failures could have been avoided if the people responsible for constructing or maintaining the equipment had possessed even an elementary knowledge of corrosion phenomena. We are convinced that there is a real need for better dissemination of the existing knowledge of metallic corrosion to designers, architects and engineers.

It is true there are a number of excellent books in English, German and Russian in which an attempt has been made to give a comprehensive coverage to corrosion in its many forms. That this is no longer possible will be obvious from the fact that some two to three thousand scientific papers on corrosion and related topics appear annually. In attempting to cover the whole field the final texts are bulky, running to over a thousand pages, and too expensive to have a wide sale outside libraries and people actively engaged in corrosion research and prevention. The engineer who is only interested in problems in a relatively narrow field would find it difficult to extract the required information from such reference books or to obtain the relevant corrosion papers which are spread thinly over a very large number of specialist journals. In addition, wide experience of corrosion and corrosion testing is necessary to apply the available information to a specific corrosion problem which will usually involve a large number of variables. In particular the enquirer could be seriously misled if the results of laboratory tests were applied indiscriminately to corrosion problems in the field.

The time has now arrived when even the specialist cannot hope to have an intimate knowledge of the whole subject and, as with other scientific disciplines, people tend to specialise in particular fields of corrosion. This has given rise to a number of shorter texts on topics such as the corrosion of iron, atmospheric corrosion, stress corrosion and cathodic protection.

Since many industries are concerned with the transport and use of water as a processing fluid or heat transfer medium it seemed to us that the corrosion of metals by waters, available in nature or treated for industrial purposes, was a suitable subject for treatment in a short monograph.

In writing this book the practical aspects of corrosion have been kept well to the fore and theoretical explanations have only been given where they are necessary for clarity and understanding. Emphasis is placed on the knowledge and experience reported on the behaviour and operation of industrial plant and equipment and much of the subject matter is capable of direct application to practical problems.

The chapters have been arranged in a logical sequence starting with the principles of corrosion, a description of the different types of water and the forms of corrosion that may be encountered. This is followed by a discussion of the corrosion behaviour of ferrous and non-ferrous metals commonly used in the construction of plant and equipment. The next three chapters deal in turn with the influence on corrosion of contact between metals, of metallurgical and mechanical factors, and of flow, temperature and heat transfer in practical systems.

The later chapters on corrosion prevention include water treatment and inhibition, the use of metallic and non-metallic coatings and the application of cathodic protection. Finally consideration is given to the importance of design and method of operation of industrial plant and also diagnosis of corrosion failures.

We hope that those readers who wish to pursue this important subject will be encouraged—and stimulated—to read the larger and more extensive books and journals with greater objectivity.

In gathering information for this book we have made an extensive survey of the whole of the world literature on corrosion and acknowledge the help we have received particularly from the large standard works written by U. R. Evans and N. D. Tomashov and those edited by H. H. Uhlig, F. Tödt, L. L. Shreir, and F. L. LaQue and H. R. Copson. Even so, it is appreciated that corrosion prevention and cure is often more of an art than a science and there are still many points of controversy; it would therefore be wrong to blame other authors for a number of opinions and conclusions in this book which are essentially our own! Much of the material is based on ideas gained from many years of experience both in applied research and in the solution of practical problems. In this we have benefited from the mutual exchange of opinions and knowledge with colleagues and would extend our thanks to Dr. W. H. J. Vernon, Dr. F. Wormwell and all those with whom we have worked.
Plates 1 to 14 (inc) and 17 to 22 (inc) are Crown copyright and published by permission.

Finally our thanks are due to Margaret (wife of H. C. K. I.) for accepting the onerous task of typing the manuscript and to Joan (wife of G. B.) for drawing the map showing the distribution of waters.

Contents

Chapter I

> Dissimilar Metals
> Metal Heterogeneities
> Surface Films
> Variations in Liquid
> Physical Conditions

> Polarisation
> Conductivity
> Relative Areas

> pH Value
> Dissolved Salts
> Dissolved Gases

Chapter II

Total Dissolved Solids
> Chlorides and Sulphates
> Carbonate and Bicarbonate
> Minor Inorganic Constituents
>> Silica
>> Iron
>> Copper
>> Lead
>> Zinc

Chapter IV

IRON AND ITS ALLOYS

Chapter VI

CORROSION AT BIMETALLIC CONTACTS

Galvanic Series
Relative Areas
Conductivity of Water

Magnesium and its Alloys
Zinc and its Alloys
Aluminium and its Alloys
Iron and its Alloys
Copper and its Alloys
Carbon

List of Tables

Table

List of Plates

List of Figures

      Seamless tubes: curve 1, 4 ml/l oxygen
                       "   2, 15 ml/l   "
                       "   3, 4-15 ml/l oxygen, pH 8
      Welded tubes: "   4, 4 ml/l oxygen
      Shaded area: Occasional tube perforations in this region.

Fig   *Page*

Introduction

Corrosion is the destruction of metals by interaction with the environment. Although the term is sometimes applied in a wider sense to the deterioration of plastics, concrete and wood, the majority of people have metals in mind when speaking of corrosion. No doubt in the future as non-metals come more and more into use their "corrosion" will receive more attention and the definition will be broadened.

Metals are used for the manufacture and construction of objects as widely different in size as razor blades and industrial plant and as widely different in purpose as washing machines and artificial satellites. In all these uses their corrosion behaviour is equal in importance to their mechanical properties. Whether it be in the atmosphere, immersed in water or buried in the ground a metal can only perform its function so long as the deterioration due to chemical interaction with its surroundings does not impair its behaviour to a significant extent.

Evidence of corrosion is all around us and we cannot go a day without being reminded of its ever-present action. In the home by the rusting of steel window frames and the presence of rust stains in the bath; in the garden by the rusting of tools inadequately cleaned, on the road by the corrosion and damage to paint on vehicles, and the obvious widespread corrosion of industrial premises.

Corrosive action by the environment is not always unwelcome as is shown by the green patina developed on church roofs, which is not only pleasing to the eye but also prevents further attack. It is not surprising that corrosion has found a place in modern art "Aetiography, " or the study of the destruction of solid bodies, is being developed as a new art form in which corrosion is controlled to produce forms of artistic merit.

Corrosion is part of a natural and inevitable cycle of events in which metals tend to return to the state whence they came. Most metals are extracted from ores with the expenditure of a considerable amount of energy and, given the environment, many metals will corrode and return to the more stable original state. Thus the corrosion products of iron in contact with water containing oxygen include lepidocrocite, γ-Fe_2O_3 . H_2O or $FeOOH$; geothite, α-Fe_2O_3 . H_2O, and magnetite, Fe_3O_4, all of which are found in the natural state. The green patina on copper has essentially the same composition as the mineral brochantite, a basic copper sulphate, $CuSO_4$. $3Cu(OH)_2$. Only the noble metals platinum, gold and silver and the "near noble" metal, copper, are found in the metallic state in nature and are very resistant to corrosion. However, instead of making adequate use of the world's supply of gold, man prefers to bury it again in underground vaults!

The amount of corrosion that may be tolerated will depend on the equipment involved. Thus in a thick-walled pipe carrying water a large amount of attack can proceed before a hole develops. On the other hand on a fine screw thread or on slip gauges the smallest particle of corrosion product can make these items useless.

In order to examine more fully the implications of corrosion we may consider a piece of metal on which corrosion is proceeding. This will first of all affect its thickness and mechanical properties and hence this must be allowed for in design calculations not only making necessary the specification of a greater thickness but also the provision for renewal of the part at some future date. The corrosion product may also contaminate the fluid in contact with the metal leading to undesirable blockages of valves and adverse effects on the industrial process. This product will also increase the roughness of the surface and provide a greater viscous drag on moving fluid and a greater expenditure of energy is required to move the fluid past the metal at the desired rate. The transfer of heat across the surface of the metal will be reduced since the thermal conductivity of the product will be considerably lower than that of the metal.

When the metal is ultimately perforated this will lead to loss of fluid or contamination between fluids on the two sides. In order to repair this leak the plant may have to be shut down, the leak detected and the defective part replaced. This will mean loss of production time as well as the cost of materials and labour in making the repair. Where high pressures or temperatures are involved corrosion action is always a danger to personnel as it may result in explosions or the escape of dangerous liquids or gases.

If more corrosion resistant material is used or certain measures are taken to reduce corrosion this will obviously increase the capital and maintenance costs respectively. In these days of high taxation corrosion prevention may not always "pay". Prevention does not increase revenue but lowers costs which may be unattractive when maintenance is directly deductible against tax.

Probably one of the worst features of industrial corrosion is the lack of appreciation by management and operating staff that any problem exists and that corrosion is costing money. There are many instances in which individuals profess to have no corrosion problems while at the same time their factories have scrap heaps and stores of replacement parts which always furnish enough evidence to contradict their views. There is also room for improvement in the collaboration and exchange of experience between users. It is all too often true that neighbouring firms using the same water supply fail to exchange information which would be to their mutual benefit in solving their corrosion problems.

It is very difficult to make a precise estimate of the cost to the national economy of corrosion, but recent calculations by Uhlig in the U.S.A. and Vernon in Great Britain both indicate that the annual cost amounts to about £12 per head of population (man, woman and child), the total bill for Great Britain being £600 million. This means that corrosion costs the average family about £1 per week as extra cost on purchases, cost of replacement of articles or in taxes. These figures will probably be equally applicable to all industrialised countries and in Australia the estimated relative expenditure is about the same. Even in India, which has not yet reached a high level of industrialisation, the corrosion bill is put at £113 million.

Corrosion is broadly classified as gaseous, atmospheric, immersed or underground depending on the environment. There are no clear-cut divisions between the last three since they all require the presence of water. In atmospheric corrosion the supply of oxygen from the air is unlimited and attack depends on the presence of water and impurities dissolved in it. In immersed corrosion in natural waters, on the other hand, attack is usually controlled by the amount of dissolved oxygen which can vary considerably. In underground corrosion all the three factors, water, salt content and oxygen vary, as well as the nature of the soil aggregate. While this book is concerned with corrosion in waters many of the principles will clearly be applicable to some extent to corrosion in other environments.

Corrosion rates may be given in a confusing variety of units. The two main classes of units are weight loss per unit area per unit time, e.g. milligrams per square decimeter per day, mdd, and penetration per unit time, e.g. mils per year, mpy. The former represents a greater loss of thickness for a light metal such as aluminium than for a heavy metal such as lead. We favour the expression of corrosion rates in mdd for three reasons:-

(1) In this form we describe exactly the quantity which has been measured since corrosion is usually measured by the overall weight loss.

(2) A day is closer than a year to the period of most laboratory tests, i.e. from a few days to several weeks. The rate of corrosive attack is rarely constant over long periods and long extrapolations from short term tests will almost invariably overestimate the rate of attack over a period of a year.

(3) There is no misleading implication that the attack is either uniform or localised. Almost invariably the attack in natural waters is localised and the only justification for the use of the term mils per year is when actual depth of pitting has been measured over a period of the order of a year. Reference is only made to mpy when such measurements have been made but elsewhere, if possible, we have qualified the data given by information on the nature of attack.

For the convenience of readers who wish to convert mdd to mpy the following approximate multiplication factors may be used:-

Magnesium	0.83
Aluminium alloys	0.53
Zinc	0.20
Iron alloys	0.19
Nickel alloys	0.16
Copper alloys	0.16
Lead	0.13

In general, iron has been used to illustrate corrosion principles because it is the most widely used metal and because its behaviour is so complex that it embraces many of the features encountered in the corrosion of other metals. Thus the confusion that might arise if we were to switch from one metal to another is avoided. As a very rough guide a corrosion rate for mild steel of less than 25 mdd is <u>good</u>, between this figure and 250 mdd the rate is <u>satisfactory</u>, but rates greater than this would be considered <u>unsatisfactory</u>.

In discussing alloys we have only used a trade name where it has become internationally accepted as the generic term for a particular class of alloy, e.g. monel. We have preferred to give the essential composition in the form illustrated by "70/30 copper-zinc" which indicates an alloy containing 70 per cent copper and 30 per cent zinc.

In a book of this nature it is impossible to avoid a certain amount of repetition particularly if we try to make each chapter reasonably complete in itself. Thus in discussing the forms of corrosion (Chap. III) there is some overlap in content with Chapter VII which deals with mechanical and metallurgical factors. We have endeavoured to place the main information on any particular topic in what we consider is the most logical place and then to refer to it elsewhere when necessary. We have not hesitated, however, to repeat certain facts if we consider it essential for the understanding and completeness of a chapter.

Chapter 1

CORROSION PRINCIPLES

ELECTROCHEMICAL NATURE OF ATTACK

A large amount of study of corrosion, particularly over the last thirty
years, has demonstrated that the primary process in the dissolution of
metals in water is electrochemical in nature, attack is basically a chemical reaction accompanied by the passage of an electric current. For this
to occur a potential difference must exist between one part of the struc-

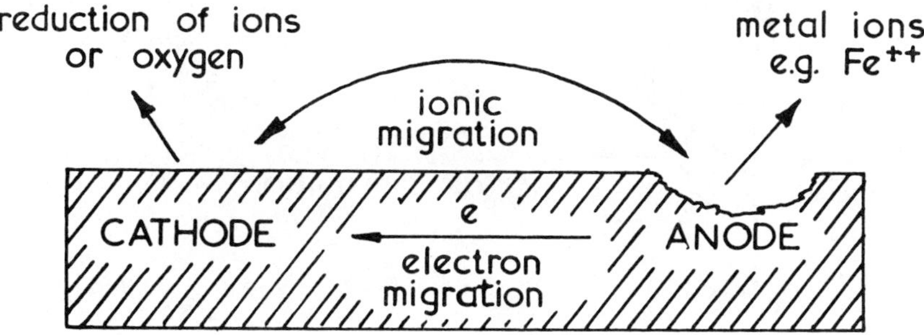

Fig. 1. SIMPLE CORROSION CELL

ture and another (Fig. 1). The primary reaction in the region at the lower
potential, the anode, is the dissolution of metal in the form of ions, e.g.

$$Fe \rightarrow Fe^{++} + 2e \text{ (electrons)}$$

The electrons liberated migrate through the metal to the part at the higher
potential, the cathode, where they are utilised in the reduction of either
ions or oxygen. The overall effect is the passage of a current through the
circuit formed by metal and solution, the current carriers being electrons
in the metal (N.B. flowing by convention in the opposite direction to
current) and dissolved ions in the solution. Positively charged cations
such as H^+ and Na^+ migrate to the cathode and anions such as OH^-, Cl^-
and SO_4^{--} to the anode.

A number of reactions is possible at the cathode (Fig. 2) including:

1. Reduction of hydrogen,

$$2H^+ + 2e \rightarrow 2H \rightarrow H_2$$

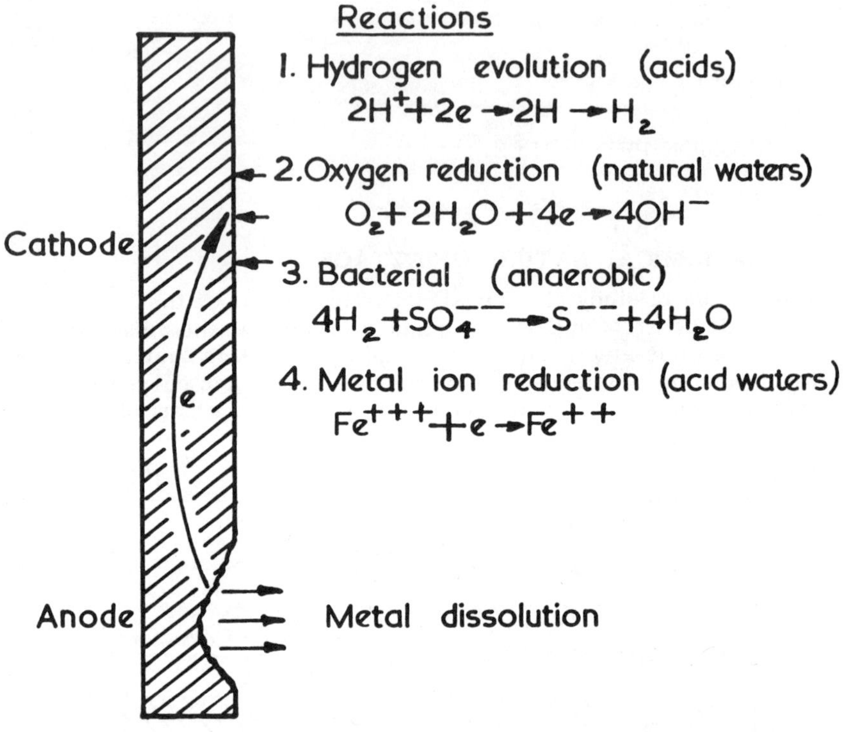

Fig. 2.   REACTIONS ON THE METAL SURFACE IN A SIMPLE CORROSION CELL

2. Reduction of oxygen:

$$O_2 + 2H_2O + 4e \rightarrow 4OH^-$$

3. Reduction of sulphate (with the aid of bacteria):

$$4H_2 + SO_4^{--} \rightarrow S^{--} + 4H_2O$$

4. Reduction of metal ions:

$$Fe^{+++} + e \rightarrow Fe^{++}$$

Hydrogen reduction is the main cathodic process in acid solutions where the metal dissolves with the simultaneous evolution of hydrogen gas. The accumulation of metal ions in the solution can lead to active participation of these ions in the corrosion process giving rise to the alternative cathodic process (reaction 4). This can occur with metals which have two valencies such as copper and iron and which can exist in solution as cupric and cuprous or ferric and ferrous ions respectively. This type of cathodic reaction is usually considered to be the cause of the corrosion of iron in acid mine waters, the ferrous iron formed at the cathode by reaction 4 being subsequently oxidised back to the ferric form by dissolved oxygen.

Reaction 2, oxygen reduction, is mostly responsible for the corrosion of metals in natural waters which have an approximately neutral reaction, i.e. are only slightly acidic or alkaline. Corrosion in such a medium is accompanied by the formation of solid product by interaction between the anodic and cathodic products, e.g.

$$Fe^{++} + 2OH^- \rightarrow Fe(OH)_2$$

When the solubility of ferrous hydroxide is reached, a white product will start to precipitate from solution. In oxygenated conditions this will be rapidly oxidised to form, firstly, ferric hydroxide,

$$4Fe(OH)_2 + O_2 + 2H_2O \rightarrow 4Fe(OH)_3$$

This is unstable and subsequently loses water to form hydrated ferric oxide, FeOOH, or Fe_2O_3 (red rust),

$$Fe(OH)_3 \rightarrow FeOOH + H_2O$$

One of two forms of hydrated ferric oxide is normally found in the corrosion product of iron in water; α-FeOOH, similar in structure to the mineral goethite, and γ-FeOOH, similar in structure to the mineral lepidocrocite or limonite. Magnetite, Fe_3O_4, haematite, α-Fe_2O_3 and maghaemite, γ-Fe_2O_3 are also formed.

In the absence of oxygen, e.g. in water from which the air has been removed, corrosion of iron can still proceed by reaction 3 in the presence of adequate dissolved sulphate and this is usually present in sufficient concentration in most natural waters. The bacteria, <u>disulpho-vibrio disulphuricans</u>, are able to use cathodic hydrogen in their living process and bring about the reduction of sulphate to form sulphide. This type of cathodic process can also occur in aerated solutions beneath any impervious corrosion product which prevents access of oxygen to the surface of the metal.

In any given circumstance more than one of the four possible reactions cited may participate in the overall cathodic process and the latter may vary with time. Thus, in acid solutions reduction of metal ions and hydrogen ions can occur, while in only slightly acid or alkaline solutions a certain amount of hydrogen is evolved even when oxygen is present. Again, in a closed vessel, corrosion can proceed by oxygen reduction until all the oxygen has been removed when sulphate-reducing bacteria can take over.

TYPES OF GALVANIC CELL

A large number of factors can give rise to variations in potential on a structure immersed in water, the galvanic cell formed giving rise to the flow of current which is necessary for corrosion. Such heterogeneities may vary in size from the sub-microscopic or atomic scale to the macroscopic when the separate anodic and cathodic areas may be seen with the naked eye. These variations may be due to differences either on the metal or liquid side of the metal/liquid interface, to the presence of surface films or to variation in physical conditions, e.g. temperature.

DISSIMILAR METALS

If we take a number of metals or alloys and immerse them in a salt solution or natural water they will, after a time, each attain a potential which is characteristic for that metal and the solution under investigation. In this way it is possible to arrange all the metals in series in such a way that, on connecting any metal to another lower in the series, current will flow from the first to the second metal (Fig. 3). The first metal is electropositive, or cathodic, to the second and such a series, known as a galvanic series, is of great value in predicting the behaviour of metals when connected together.

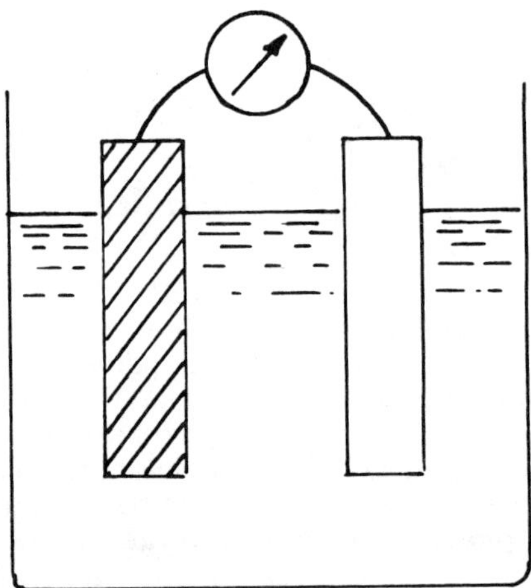

Fig. 3.   CORROSION CELL FORMED BY TWO DISSIMILAR
          METALS

Since the most serious corrosion failures with mixed-metal couples are encountered in conducting media, i.e. where the resistance path through the water is low, the only natural water in which any extensive series of metals has been studied is sea water, which conveniently maintains a reasonably constant composition throughout the oceans and seas of the world. Like potentials have been recorded in 3 per cent sodium chloride solution which has a similar chloride content to that of sea water. The corrosion resistance of metals generally decreases as we move down the series from the noble or cathodic end where we find platinum and gold, to the base or anodic end, where magnesium is situated. In general, movement of the water will displace the potential in a positive, more noble, direction.

There are many practical examples of bimetallic couples, e.g. a copper pipe connected to a galvanised tank and a bronze propeller in contact with the steel hull of a ship (Chap. VI).

A composite series of potential values of metals and common alloys derived from a number of sources is shown in Fig. 4. Here the potentials have been expressed on an absolute scale relative to a hydrogen electrode in the same solution. This electrode, which consists of a platinum sheet, gauze or wire over which hydrogen is passed at a pressure of one atmosphere, is taken as being at zero potential when the activity of hydrogen ions in the solution is one gram-ion per litre, approximately 1.2 N. hydrochloric acid. Potentials may be measured directly against such an electrode, or more conveniently measured against some other electrode of known potential relative to the hydrogen electrode (Tab. 1). Potentials referred to the normal hydrogen electrode, i.e. v NHE, are said to be on the <u>hydrogen scale</u>.

The potential of rusting iron is usually about 0.65 volts negative to the saturated calomel electrode. On the hydrogen scale this will be $-0.65 + 0.242$ volts or -0.41 volts.

TABLE 1

POTENTIALS OF ELECTRODES

Name	Cell	Potential at 25°C. Volts	Temperature variation. mV per °C
0.1N calomel	$Hg/Hg_2Cl_2/KCl$ (0.1N)	0.334	-0.07
1.0N calomel	$Hg/Hg_2Cl_2/KCl$ (1.0N)	0.280	-0.24
Saturated calomel	$Hg/Hg_2Cl_2/KCl$ (sat.)	0.242	-0.76
Copper sulphate	$Cu/CuSO_4/CuSO_4$ (sat.)	0.318	0.90
Silver chloride	$Ag/AgCl/KCl$ (0.1N)	0.288	-0.65
Silver chloride	$Ag/AgCl/KCl$ (1.0N)	0.222	0.60
Silver chloride	$Ag/AgCl/KCl$ (sat.)	0.225	0.60
Silver chloride	$Ag/AgCl/Sea$ water	0.250	—

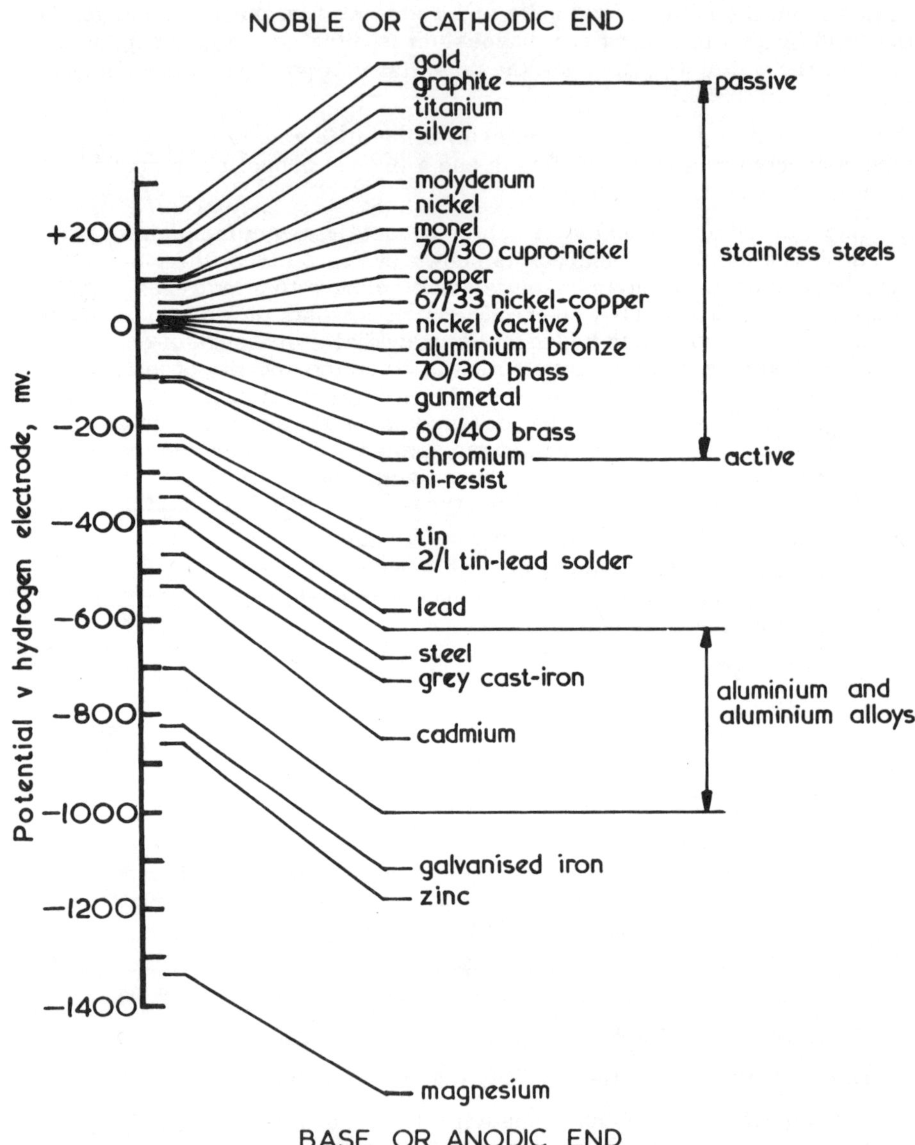

Fig. 4. PRACTICAL GALVANIC SERIES OF METALS AND ALLOYS

METAL HETEROGENEITIES

In the model of a perfect metal surface we can imagine the individual metal ions as being distributed over the surface in a regular array with each ion in an identical environment, (Fig. 5A). Thus the metal atoms will all be in an identical energy state and no atom will have any greater or less tendency to go into solution than any other. In other words the probabilities for all the metal ions to go into solution will be the same and no corrosion can take place. In the practical case conditions will be otherwise. At any temperature other than absolute zero, the ions in the crystal will be in thermal vibration and at any moment the energy states of the atoms will differ. Furthermore no surface, however carefully pre-pared, will exist as a plane on the atomic scale, (Fig. 5B). Edges, E, and corners, C, will exist at the boundary of incomplete lattice planes and atoms situated there will have a greater tendency to go into solution than atoms in the middle, M, of a completed plane since the bonds to be broken will be fewer in number. We should expect that the stability will increase in the series:- C → E → M → I (atoms within the metal). Removal of these atoms by dissolution at these anodic points will in turn produce new anodic sites. Thus on the surface of even the purest metal with only one plane exposed to a solution there is the possibility of anodic and cathodic processes occurring at sites which vary in position with time.

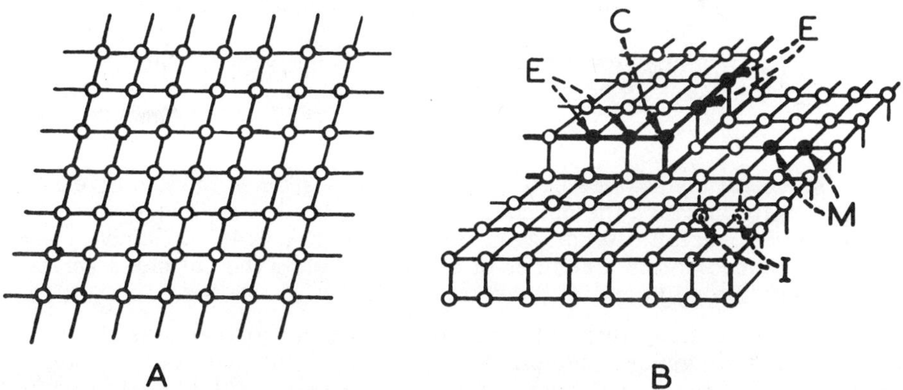

Fig. 5.   REPRESENTATION OF, (A) IDEAL AND (B) REAL, METAL LATTICE

In many cases, however, the atoms will not be identical. Even in a single-phase alloy the neighbouring metal atoms may be different. Micro-cells will be set up according to the positions of the constituent metals in the galvanic series. In brasses where the zinc is in solid solution in the copper, it is possible for zinc to dissolve leaving behind a spongy matrix of copper. The initiation of attack is probably due to such variations on the atomic scale but subsequent attack will no doubt be controlled by the macro-couple copper/brass.

Most industrial materials are polycrystalline with the grains or crystallites exhibiting different facets to the solution. The density of atoms will vary from grain to grain and at the same time the stability and, what amounts to the same thing, the potential, will vary as we move across the metal surface, micro-cell 1-2, (Fig. 6). Certain planes, often those with the larger density of metal atoms, will be the more stable. Other planes will tend to be attacked in such a way as to reveal such planes. This anisotropy is made use of in metallurgical examination when acid etching of highly polished metal surfaces results in preferential etching of grains with particular orientations. This difference in orientation can also be observed on galvanised iron on which the zinc crystallite size is particularly large.

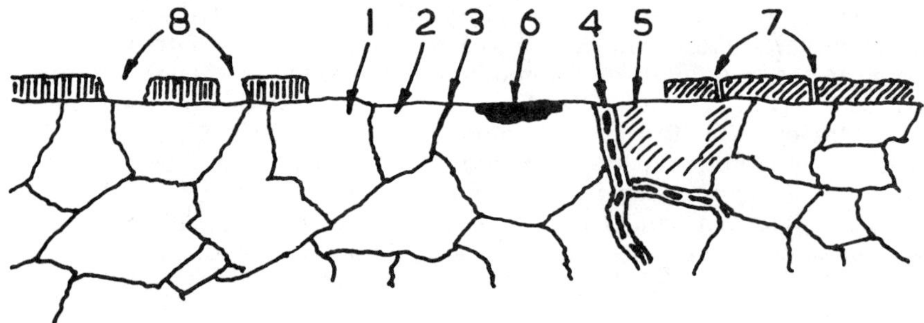

Fig. 6.   VARIETIES OF MICRO-CELLS FORMED ON TYPICAL METAL SURFACE

The existence of boundaries between the various grains can also give rise to micro-cells in which the boundary is usually the anode (Fig. 6.3). Furthermore, precipitation in grain boundaries (Fig. 6.4) may cause intercrystalline corrosion in a number of alloys including the stainless steels and certain aluminium and magnesium alloys.

Even in solid solution alloys, differences in composition can exist from place to place. In general, the parts of the alloy which contain a greater concentration of the more noble phase, i.e. component with the more positive potential, will be cathodic to the rest of the surface (Fig. 6.5). Such conditions can result in the increased corrosion of aluminium-zinc alloys in aqueous solutions.

Certain macro- or micro-inclusions, which may be metallic or non-metallic but must be electrically conducting, can produce galvanic cells (Fig. 6.6). Contacting materials or inclusions with more positive potentials will invariably act as cathodes but the degree to which they are able to stimulate attack depends on a number of factors and will be considered later. The most striking and obvious example is the corrosion of a metal in contact with a more electropositive metal. Another example of this type of attack is the accelerating influence of carbon films on the corrosion of copper tubes in supply waters.

Galvanic cells with a single material arise from the different potentials
of areas of unequal stress and deformation. The more stressed parts are
usually anodic and corrode more readily. Many factors can cause these variations, e.g. strains set up by rivets and bolts or external stresses. This
topic, a very important one indeed, will be discussed at greater length in
Chap. VII. Here it is only necessary to cite a few examples such as corrosion in boilers, cracking of brass, corrosion at bends in iron and corrosion of the heads of rivets to appreciate the magnitude and importance
of this subject.

SURFACE FILMS

The cases discussed are those in which the solution has direct access to
the metal surface. In natural waters the conditions will generally differ
from this and will be complicated by the presence of an oxide film or a
layer of corrosion product on the metal surface. This will tend to throw
the emphasis on the variability of the layer which develops between the
metal and the solution. Thus when a protective film, e.g. the oxide film on
aluminium, begins to break down attack can proceed at the micro-pores
(Fig. 6. 7).

On a macro-scale, discontinuous films of oxide on the metal surface will
act in such a way as to render the part not covered by the oxide layer
anodic (Fig. 6. 8). The discontinuities in the mill-scale on iron or steel
are often responsible for severe corrosion of the basis metal. In all too
many cases, lack of appreciation of this fact has resulted in serious and
quite easily avoidable deterioration of iron structures.

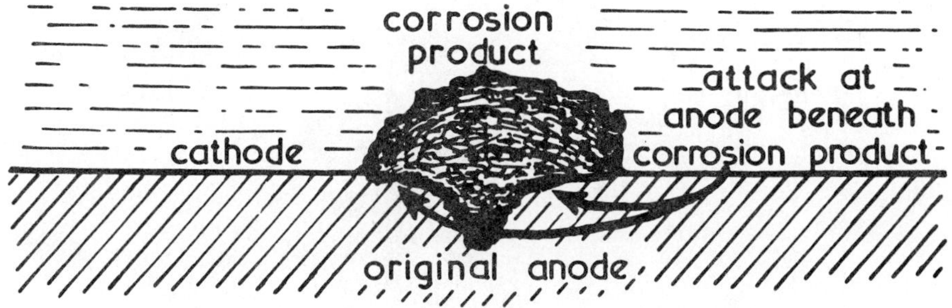

Fig. 7　INFLUENCE OF CORROSION PRODUCT ON DISTRIBUTION OF
　　　　ATTACK

Whatever the initial cause of corrosion, the presence of corrosion product
on the surface has a marked influence on the distribution of attack (Fig. 7).
The parts of the surface beneath such products are generally more anodic.
A good example of this is the influence of rust on the subsequent corrosion of iron when galvanic cells are produced by variation in accessibility
of the electrolyte and oxygen to different parts of the surface. A discussion of this will follow.

VARIATIONS IN LIQUID

In considering variations in the liquid phase we are solely concerned with the solution in contact with the metal surface or with regions of the solution close enough to have an influence on the corrosion processes occurring there. Such galvanic cells can arise due to variations in the concentration of metal ion, neutral salt, hydrogen ion, i.e. pH value, oxygen or oxidants. A metal in a solution of its own ions will have the higher potential where the ionic concentration is greatest (Fig. 8.1). Parts of the metal in contact with the lower concentration of its ions will consequently be anodic. Such conditions will occur with copper in flowing water when attack takes place at the area from which copper ions are removed most readily, i.e. where the flow is fastest.

Differences in the concentration of neutral salts can also produce galvanic cells (Fig. 8.2). In the case of an aggressive ion such as chloride the part of the metal in contact with the more concentrated solution will be anodic

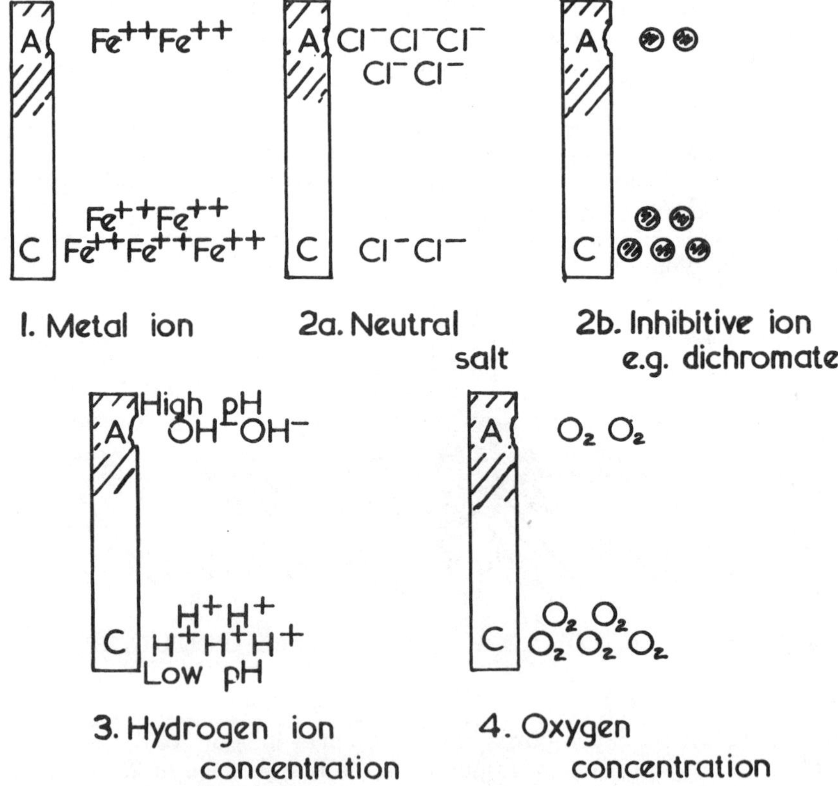

Fig. 8. GALVANIC CELLS FORMED BY VARIATIONS IN CONCENTRATION

A-anode   C-cathode

(Fig. 8. 2a). In the case of a salt with inhibitive properties such as sodium dichromate, on the other hand, the higher concentration of salt will render the surface cathodic (Fig. 8.2b). Because of the large difference in chloride concentration between fresh water and sea water, localised attack on ships and structures may be caused by chloride variations at the mouths of rivers where the fresh water flows into the sea.

Differences in hydrogen ion concentration or pH are also a possible source of potential differences with the parts of the metal in contact with the higher pH value being anodic to the remainder (Fig. 8.3). In the activation of a metal from the passive state the reverse may be the case.

Variations in oxygen concentration act in such a way as to make the metal in contact with the solution of lowest oxygen concentration the anode (Fig. 8. 4). Other oxidants will act in a similar manner. Many cases exist in which the attack can be attributed to this source, one of the most common being preferential attack in narrow cracks and crevices.

Even if these macro-variations in concentration are absent, submicroscopic heterogeneities can arise as a consequence of the thermal motion of the molecules. This can mean that alternating anodic and cathodic processes can occur even on a surface where every precaution has been taken to make the surface region homogeneous.

PHYSICAL CONDITIONS

Possible variations in physical conditions include differences in temperature, stray currents and water flow. In the case of temperature differences the hotter part of the metal is usually the anode, e.g. iron in dilute chloride (Fig. 9). In sulphate, silver behaves similarly but on copper and lead the hot part is the cathode. Such a basic cause of corrosion can obviously arise in any heat-exchange equipment; boilers and refrigerating equipment are extreme examples.

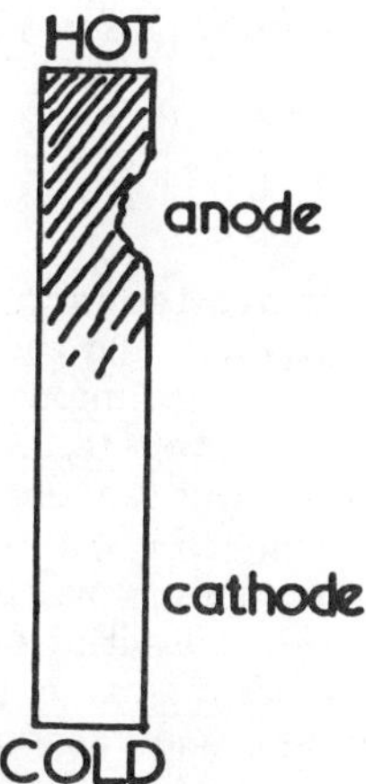

Fig. 9.   GALVANIC CELL CAUSED BY DIFFERENCE IN TEMPERATURE

Anodic and cathodic areas will arise on different parts of a metal situated in an external electric field (Fig. 10). Since the conductivity of the metal will invariably be greater than that of the surrounding water any current passed will choose the path through the metal in preference to the water The region where the positive current leaves the metal and goes into the electrolyte will be the anode while the point of entry will be the cathode. This is a regrettably frequent cause of corrosion in the case of buried pipes and in installing cathodic protection it is important that adequate precautions are taken. A similar cause of increased attack can arise owing to careless earthing of electrical installations in ships or of welding equipment in shipyards.

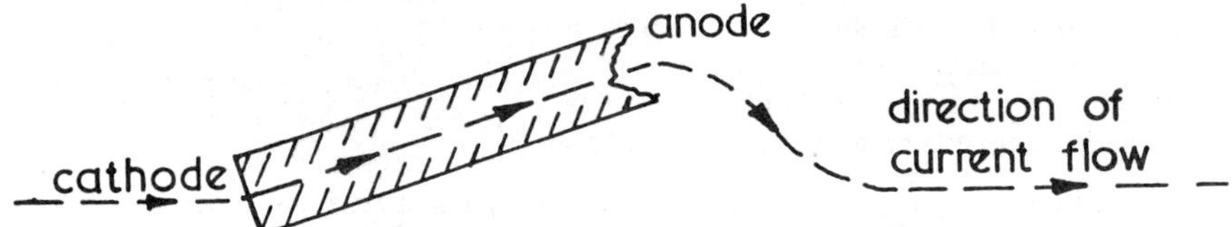

Fig. 10.   GALVANIC CELL CAUSED BY EXTERNAL
ELECTRIC FIELD

Flow has a similar effect on the cathodic process to an increase in concentration of reactants at the metal surface. It may also affect the anodic process by facilitating removal of corrosion products.

As we have seen there is a large number of ways in which galvanic cells are created. In a given situation the cause of attack may be traced to one or more of these but as corrosion proceeds the factors responsible for the attack may change. Thus the accumulation of corrosion product around an area of localised attack will produce a spread of attack because of the formation of concentration cells. (See Fig. 7).

INTENSITY OF ATTACK

All we have considered up to now are the factors that determine the tendency for a structure to be attacked and the factors determining the part of the structure which will undergo maximum attack. In practice we are concerned with the amount and intensity of attack. It would be natural, but very erroneous, to assume that the tendency for attack to take place has any direct or predictable connection with the amount of corrosion that actually occurs. The rate and degree of localisation of corrosion is not only controlled by the difference in potential on open circuit between the anode and cathode but also their polarisation, the conductivity of the water, and the relative areas of the anodic and cathodic metal. In other words we must consider not only the rates of the anodic and cathodic reactions but also the limitations imposed by the electrical resistance of the conducting path through the solution.

POLARISATION

In practice the anodes and cathodes of the micro-cells, formed in the
ways outlined, or the macro-cells, formed by the juxtaposition of two different metals, will be short-circuited through the metal and the measured
potential for the whole assembly will be a compromise potential. The
potential of the cathode will be displaced, or polarised, towards that of the
anode and vice versa, but small potential differences will still be observable if the reference electrode is brought close enough to the metal surface. The compromise potential may be close to the open-circuit anode or
cathode potential or intermediate between the two depending on whether
the corrosion process is under cathodic, anodic or mixed control.

As we have already seen the anodic process consists of the dissolution of
metal ions. This is a very fast reaction and since the concentration of
metal ions in the solution will usually be kept low by precipitation of corrosion product, the anodic reaction is only rate controlling with "near
noble" metals such as copper or with metals in the passive state (see
later) where the diffusion of ions through an oxide layer is slow.

In the majority of cases the current furnished by a galvanic cell in natural
water will be controlled by the oxygen reduction reaction, i.e. will be
under cathodic control. The controlling process may be the rate of transport of oxygen to the surface, diffusion control, the rate of reaction at the
surface, activation control, or the resistance between the two metals
through the solution, resistance control. This may be illustrated by considering the cells formed by copper-aluminium and aluminium-stainless
steel couples which on open circuit show a similar difference in potential.
Copper is an efficient cathode and oxygen is readily reduced on its surface:
it has a low oxygen overvoltage. In this case the interaction will depend on
the diffusion of oxygen to the copper surface and the overall effect will be
to produce pronounced stimulation of attack on aluminium. In well-aerated
solutions, and in moving waters, such stimulation may be disastrous.

Oxygen is not readily reduced on an alloy such as stainless steel which
has a passive film on its surface, i.e. having a high oxygen overvoltage. In
this case the stainless steel will be readily polarised to the aluminium
potential and the galvanic effect will be small or negligible.

CONDUCTIVITY

Another important factor in assessing the behaviour of a galvanic cell is
the conductivity of the liquid. In a good conductor such as sea water the
distant parts of the structure which are cathodic to some other area can
play an effective part in the cathodic reaction but if the liquid is of poor
conductivity, e.g. a supply water or condensate water, the flow of current
will be limited to the immediate areas of contact between the two areas.

This variation may be shown from a consideration of the behaviour of
galvanised steel. At any breaks in the zinc coating the potential of the exposed steel is polarised to a more negative potential at which ferrous ions
can no longer leave the metal, i.e. the steel is cathodically protected.
Oxygen reduction takes place on the iron cathode and increases the cor-

rosion of the zinc anode. If a large area of steel is exposed, however, the degree of protection will depend on the conductivity of the solution. In distilled water the resistance through the solution will be high and only the iron immediately next to the zinc coating will be protected. In a brackish water or sea water, on the other hand, protection will extend over the whole area.

If there is only a thin film of water or moisture in contact with the metal surface the effect will be similar. Protection of the iron will here be restricted to the area immediately adjoining the junction irrespective of the conductivity of the liquid, owing to the high resistance of the liquid path to more distant parts.

RELATIVE AREAS

This introduces us naturally to the influence of the relative areas of the cathodic metal and the anodic metal. It is clear from what has already been said on the role of the conductivity of the water that in waters relatively low in dissolved solids the area of the cathode will be unimportant since attack on the anode is restricted to the area immediately next to the bimetal junction.

On the other hand in a conductive solution such as sea water, providing the cathodic reaction is unrestricted by polarisation, the attack will be controlled by the relative areas of the anode and cathode. In the case of the aluminium-copper couple already discussed the amount of attack on the aluminium will be proportional to the area of the copper. An aluminium rivet in a more noble metal such as copper would suffer intense attack but a copper rivet in a large sheet of aluminium, although undesirable, would be less disastrous.

WATER COMPOSITION

The influence of the type of water will vary according to its conductivity, already discussed pH value, hydrogen ion concentration and composition.

pH VALUE

The pH value of solution may be considered for all practical purposes as:-

$$\text{pH} = - \log \left[H^+ \right] \, *$$

the negative value of the logarithm of the hydrogen ion concentration. This gives pure water a pH value of 7.0 at 25°C, acid solutions having pH values less than 7.0 and alkali solutions greater than 7.0. Other values apply at higher temperatures. Hydrogen ion concentration, or pH value,

---

* Throughout this book, square brackets refer to the concentration in gram-ions or gram-molecules per litre. Strictly speaking activity should be used but in dilute solutions such as natural waters the difference between activity and concentration can be neglected.

influences the corrosion rate in a varied manner depending on whether
the metal is <u>noble</u> or whether its oxide is soluble in acid or both acid and
alkali. Noble metals such as platinum and gold are stable in both acid and
alkaline solutions and their corrosion behaviour is independent of pH
(Fig. 11.1). At the other end of the scale we have metals such as zinc and
aluminium with <u>amphoteric oxides</u> which are soluble in both acid and
alkaline solutions forming Zn^{++} and Al^{+++} or ZnO_2^{--} (zincate ion) or
AlO_2^- (aluminate ion) respectively. Such metals have a parabolic dependence of rate of corrosion on pH dissolving rapidly in both acids and
alkalis (Fig. 11.2). These metals have a characteristic pH value at which
the corrosion rate is a minimum which is 6.5 for aluminium, 8.0 for lead,
8.5 for tin and 11.5 for zinc.

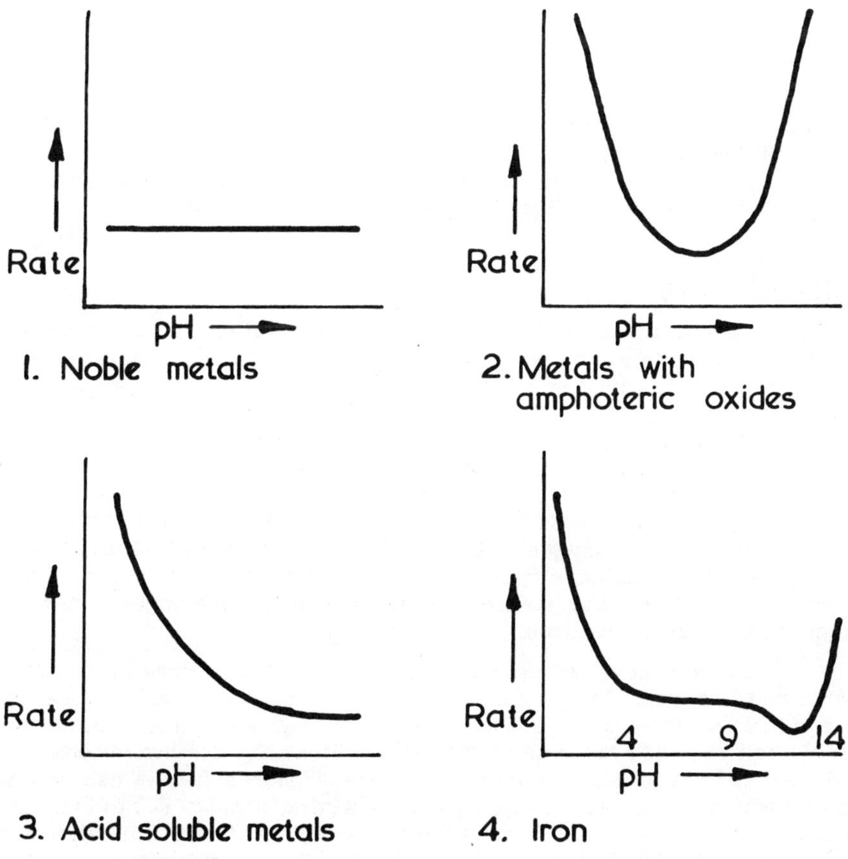

Fig. 11.   VARIATION IN CORROSION RATE WITH pH VALUE

Most metals have oxides which are soluble in acids but insoluble in alkalis and a pH behaviour shown in Fig. 11.3. Nickel, copper, cobalt, chro-

mium, manganese, cadmium and magnesium belong to this group. To a
certain extent iron falls into this class, the corrosion falling with increase
in pH to a limiting value between pH 4.0 and pH 9.0 where the rate of
attack is sensibly constant (Fig. 11.4). On further increase in alkalinity
the rate of attack falls to a minimum at pH 12.0 and subsequently increases. Iron reacts with strong caustic alkali with the formation of
sodium ferroate, Na_2FeO_2:-

$$Fe + 4OH \rightarrow [FeO_2]^{--} + 2H_2O + 2e$$

$$Fe + 2NaOH \rightarrow Na_2FeO_2 + H_2$$

This reaction is followed, to a limited extent, by the formation of sodium
ferrite, $NaFeO_2$,

$$[FeO_2]^{--} \rightarrow [FeO_2]^{-} + e$$

Thus in both strong acid and strong alkali cathodic liberation of hydrogen
occurs but in near-neutral water the reduction of oxygen is the main
cathodic reaction. In reality, therefore iron is intermediate between the
classes represented by Figs. 11.2 and 11.3.

DISSOLVED SALTS

The influence of dissolved materials in natural waters is determined not
only by the concentration of dissolved salts but also, more importantly,
by the type of dissolved salt. We may distinguish between aggressive ions
such as chloride, Cl^-, and sulphate, SO_4^{--}, and materials such as carbonate and bicarbonate, CO_3^{--} and HCO_3^{-} and calcium, Ca^{++}, which have
inhibitive properties, i.e. are capable of restraining corrosion. In effect
all types of ions have to some extent both properties according to their
concentration. Thus any very soluble salt has inhibitive properties in
that the solubility of oxygen is lower in concentrated salt solutions and
hence the rate of cathodic reduction of oxygen is reduced. On the other
hand ions with true inhibitive properties are, in general, operative above
a low critical concentration.

The specific action of any aggressive ions is to break down protective
films or to prevent their formation. These ions, in particular chlorides,
are able to destroy passive films formed on metals such as stainless
steels and increase the rate of attack considerably. Inhibitive ions such as
calcium and carbonate can lead to the formation of a film of calcium carbonate which has a restraining action on further attack. Such films can
arise in one of two ways. If the solution is supersaturated in calcium bicarbonate, deposition of carbonate will take place spontaneously on any
surface in contact with such a water. On the other hand deposition can
occur from a solution in equilibrium, just saturated or even below saturation because the conditions are disturbed by the alkali formed at the
cathode on oxygen reduction. This equilibrium will be discussed in
Chapter II.

DISSOLVED GASES

In considering dissolved gas the main attention is focussed on carbon dioxide and oxygen. Carbon dioxide in water lowers the pH value and by virtue of this fact alone can stimulate intensive attack, e.g. on iron in steam condensate. Furthermore free carbon dioxide, i.e. dissolved gas not bound in the bicarbonate or carbonate ion, renders a water aggressive. The reaction leading to the deposition of carbonate scale is reversed and any existing scale may be dissolved.

Oxygen plays a dual role both as cathodic depolariser and an anodic polariser or passivator. Within a certain range of concentration the rate of attack is dependent on the oxygen concentration. Above a certain concentration it may act as an anodic passivator on certain metals leading ultimately to the reduction of corrosion to zero or a negligibly small value. In order to understand the latter effect we must understand something of the <u>active and passive state</u> of metals.

Almost all metals with the possible exception of gold become covered with a film of oxide on exposure to dry air or oxygen. When immersed in water the stability of such a film will depend on its composition and the concentration of oxygen or other oxidising agents. The film is in <u>dynamic equilibrium</u> in regard to the solution and its stability or breakdown depends on the repair or lack of repair of the points where the film breaks down. Many metals such as stainless steel are <u>passive</u> in water and under anodic control provided that the oxygen concentration is normal. If the oxygen concentration is very low or access is poor, e.g. in crevices, such materials become <u>active</u> and proceed to dissolve. Iron behaves in a similar manner although the oxygen supply to the surface must be abnormally high for it to exist in the passive state. The rate of attack is then controlled by diffusion of metal ions through the thin film, i.e. <u>anodic control</u>.

Chapter II

TYPES OF WATER

Chemically the term "water" designates a compound of hydrogen and
oxygen which exists at ordinary temperatures in the liquid state. However,
since it is an excellent solvent, natural waters always contain various
amounts of materials dissolved either from the atmosphere or from the
ground through which they percolate. The constituents in natural or
treated waters which are significant to the corrosion engineer differ from
those which are important to the water engineer responsible for a public
supply. Waters suitable for human consumption must be palatable and
non-toxic while for laundry purposes the softer the water the more favourably it is regarded. However the waters used in industry show a much
wider variation in composition and properties.

The water used in industrial plant is dictated by geographic situation and
economic considerations and the source of supply is chosen which gives
the required volume at the lowest cost. Thus it is often necessary to use
a cheap water which is aggressive from a corrosion point of view in preference to a public supply which is more expensive. Sea water and brackish polluted estuarine waters are often used when these are readily available. Even when a non-corrosive hard water is available, for technical
reasons procedures such as softening or purification which increase the
corrosivity of the water may be necessary.

Consequently the corrosion engineer must assess the quality of a water
from an entirely different standpoint. The amount and nature of the dissolved solids, gases and contaminants are extremely important but only
certain bacteria are of any significance.

In order to make any evaluation of the probable corrosive action of a
water in an industrial installation it is essential to have adequate knowledge of the physical and chemical properties of the water. Not only must
factors such as temperature and flow velocity be known but also the important dissolved constituents in the water, the way in which they are
likely to vary during processing and, most important, possible ways in
which the water quality can be economically improved in order to reduce
corrosion to a minimum. The possible direct or secondary influence of
pollution and the presence of various micro-organisms must also be considered.

DISSOLVED SALTS

The main constituent ions in natural waters are positively charged cations
such as calcium, Ca^{++}, magnesium, Mg^{++}, sodium, Na^+, and hydrogen H^+,

and negatively charged anions such as chloride, Cl^-, sulphate, SO_4^{--}, bicarbonate, HCO_3^-, carbonate, CO_3^{--}, and hydroxide, OH^-. Since the total electric charge on the cations is balanced by that on the anions the water carries no overall charge and may be regarded as a solution of a number of salts.

TOTAL DISSOLVED SOLIDS

The total dissolved solids, abbreviated TDS, may be determined directly by evaporating to dryness, the final drying usually being carried out at 180°C, or estimated to a sufficient degree of accuracy from the electrical conductivity of the water. The total dissolved solids, in ppm, is close to one-fifteenth of the conductivity, in reciprocal ohms.

99 per cent of the waters in the British Isles have a TDS below 550 ppm (Fig. 12).

CHLORIDES AND SULPHATES

These make up the bulk of the corrosive salts present in most waters. In general the amount of dissolved chloride is greater than the amount of sulphate and only in certain highly-mineralised waters is the sulphate predominant.

Chloride is often taken as an index of the corrosive potential of the water. It varies over very wide limits indeed, from the traces found in unpolluted rain water to the high concentration found in sea water. Some of the purest land waters have chloride contents of 5 ppm or less and most public supply waters contain less than 25 ppm (Fig. 12.).

Larger amounts of chloride are derived either from a particular geological strata or by pollution of rivers by sewage and industrial effluents. For example the 10 to 20 ppm present in the unpolluted rivers or streams may be increased in the average domestic sewerage to about 100 ppm. Still greater amounts are often found in underground sources particularly in desert or in coastal regions where there may be infiltration of sea water which has the high chloride concentration of about 35,000 ppm. This contamination may be only moderate and tolerable but the chloride can rise to a troublesome level by over-pumping. Very wide variations can occur in river estuaries particularly those which are tidal such as the Severn. Waters high in chloride are often referred to as "brackish" and have an unpalatable taste. The wide range of chloride and sulphate concentration in natural waters is indicated in Table 2.

CARBONATE AND BICARBONATE

These constitute the bulk of the dissolved salts in natural waters. They are closely linked with the <u>carbon dioxide</u> and <u>calcium</u> content of the water and will be fully discussed together later. In the area roughly to the south-east of a line drawn from Torquay to South Shields calcium bicarbonate predominates and is responsible for the greater hardness of these waters. To the north of this line underground waters flowing through the older geological formations are softer but contain larger amounts of mineral constituents and sulphate and chloride (Fig. 13.).

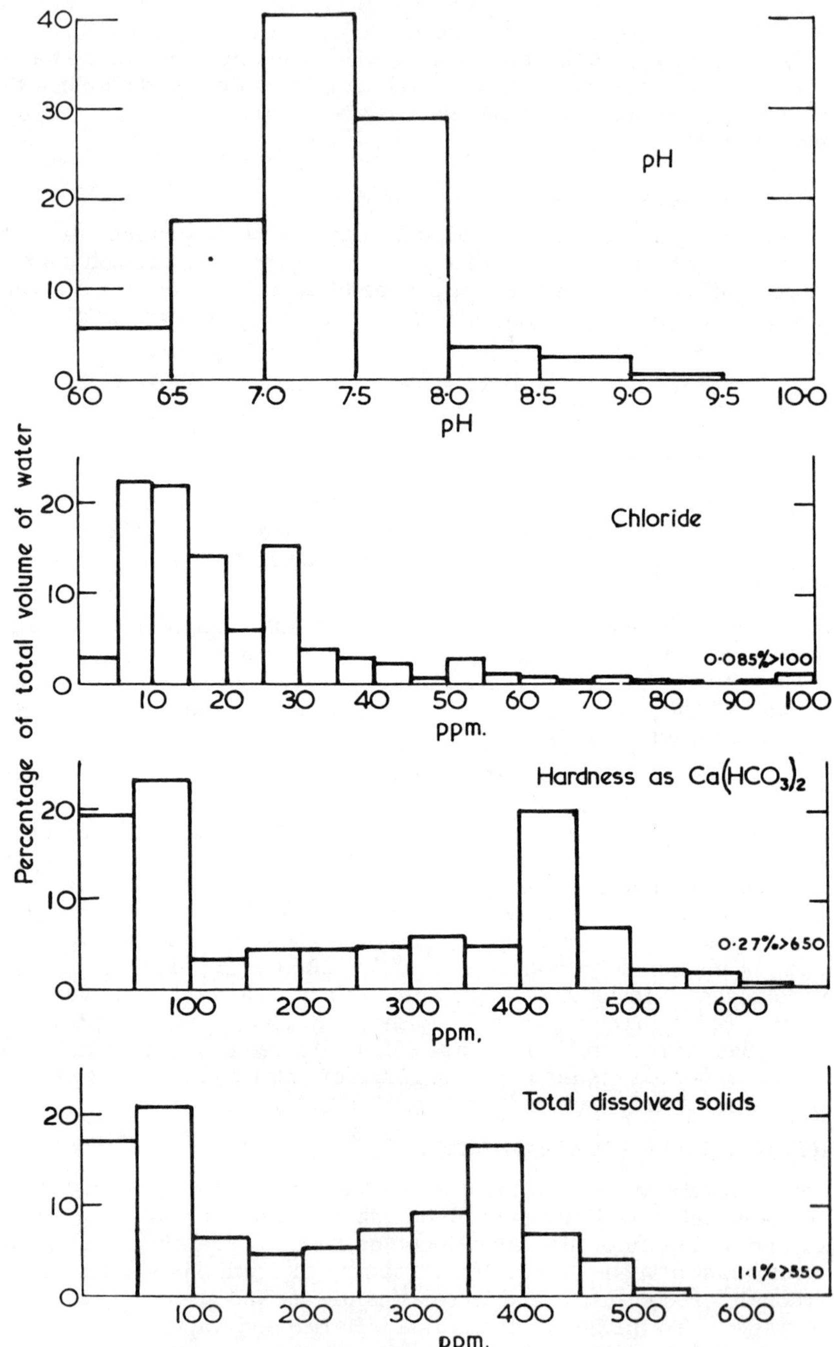

Fig. 12.   DISTRIBUTION OF DISSOLVED CONSTITUENTS IN
BRITISH WATERS

TABLE 2

COMPOSITIONS OF NATURAL WATERS ARRANGED IN INCREASING CHLORIDE CONCENTRATION

(Concentration in ppm)

Type	Source	Total dissolved solids	Cl^-	SO_4^{--}	Ca^{++}	Mg^{++}	Alkalinity $(CaCO_3)$	pH	Hardness $(CaCO_3)$ Total	Hardness $(CaCO_3)$ Temporary	Free CO_2
Reservoir	Vaich Res., Scot.	37	Trace	17	3	<1	—	6.2	—	—	—
Bay	Shawinigan Falls, Quebec	34	2	0	—	—	—	6.8	15	—	—
Lake	Lake Vyrnwy, Wales	40	7	5	—	—	8	7.0	16	—	3
Mains	Bristol	283	14	31	105	10	—	7.6	210	154	5
Mains	Teddington	360	51	32	118	20	200	7.5-8.0	320	240	5
Mains	Kuwait	336	108	48	37	10	—	9.3	—	—	—
Deep well	Reading	511	117	39	42	29	—	—	233	216	(+HCO_3) 152
Borehole	New Windsor	870	215	—	—	—	265	7.7	195	195	7
Artesian well	New Windsor	1,250	320	—	—	—	271	—	700	150	—
Mains	Bledington, Oxon	2,968	540	1,080	35	12	—	8.1	165	165	—
Mains	Benghazi	—	540	112	82	64	210	7.8	470	120	—
Mains	Aden	1,860	582	423	103	113	160	7.8	715	160	4
Mains	Malta	1,525	681	65	142	37	—	—	393	—	—
Well	Damman (Saudi Arabia)	2,656	1,041	489	241	89	—	—	—	—	—
Borehole	Harwich	2,577	1,132	136	—	—	—	7.8	602	305	—
Well	Awali, Bahrein	15,000	7,310	561	688	309	160	7.1	2,990	—	—
Sea	Average figure	34,800	19,500	2,380	380	1,250	—	8.2	—	—	—

Metamorphic
Rocks
Carboniferous
Silurian
Carboni.

1 – Up to 50 ppm – Very soft
2 – 50 to 100 ,, – Moderately soft
3 – 100 to 150 ,, – Slightly hard
4 – 150 to 200 ,, – Moderately hard
5 – 200 to 300 ,, – Hard
6 – Over 300 ,, – Very hard

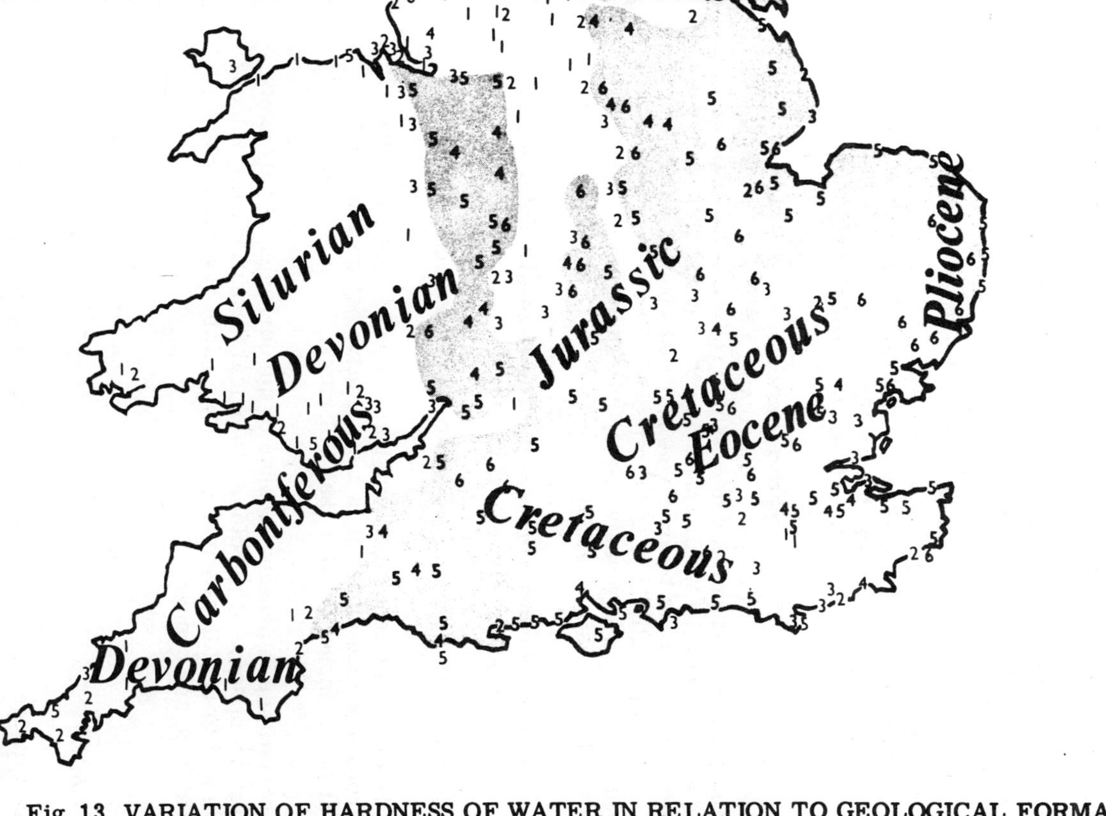

Fig. 13 VARIATION OF HARDNESS OF WATER IN RELATION TO GEOLOGICAL FORMATIONS

MINOR INORGANIC CONSTITUENTS

Among the minor inorganic constituents present we have silica and traces of certain heavy metals which are indicative of the corrosive nature of the water or its toxicity.

Silica. As a result of the disintegration of rocks and derivation from micro-organisms, particularly diatoms, silica is often present in waters usually as metasilicic acid, $(H_2SiO_3)_n$. The concentration varies in natural waters from a trace to over 75 ppm SiO_2 but the amount is usually in the range 5 to 30 ppm.

Silicates have certain inhibitive properties and are added to soft waters to reduce corrosion and are used as conditioning agents in low-pressure boilers. In high-pressure steam-raising equipment silica is undesirable since even at small concentrations it forms hard incrustations in the boilers and, through carry-over in the steam, on turbine blades.

Iron. Iron is sometimes present in natural waters, often as ferrous carbonate, at concentrations up to 20 ppm. On coming into contact with the air it is oxidised and rust is precipitated. This "red water" causes unsightly stains and renders the water unsuitable for domestic and many industrial uses. Iron starts to become a nuisance at about 0.2 to 0.3 ppm and larger amounts generally have to be removed. These ferruginous waters are indicative of an aggressive water containing larger amounts of free carbon dioxide than usual (see later).

Copper. Copper is not normally present in natural waters and when present in tap water it is usually derived from copper pipes and storage cylinders. Very small amounts are capable of stimulating attack on aluminium, to a less extent on zinc and to some degree on iron.

Lead. Lead is rarely present in the original raw water but can arise from the corrosion of lead pipes. It is a cumulative poison and should not be present in an amount greater than 0.1 ppm.

Zinc. Zinc sometimes is present because of the corrosion of galvanised iron. Mine waters in some of the older geological formations, e.g. mines in Derbyshire and Cornwall, are acid and may contain several parts per million of lead, zinc and copper.

DISSOLVED GASES

In its passage through the air, water dissolves nitrogen, oxygen and carbon dioxide and, in polluted atmospheres, small amounts of hydrogen sulphide, sulphur dioxide and ammonia. Further amounts of gases derived from decaying vegetation are dissolved during its passage through the ground.

NITROGEN

The nitrogen content of the water has little direct effect on the corrosion reaction but bubbles of gas can give rise to impingement or cavitation attack.

OXYGEN

Most public supplies are well oxygenated with an oxygen content of 2 to 8 ppm at ordinary temperatures. The solubility decreases with rise in temperature and is virtually zero at the boiling point. In a closed system, however, the solubility is dependent on the partial pressure of oxygen in the gas space and the temperature. For a given partial pressure of oxygen the amount dissolved decreases with rise in temperature to just above 100°C and then increases, the solubility at 200°C being similar to that at 25°C.

Deoxygenation often occurs in "dead ends" and stagnant zones of distribution schemes. In certain reaches of polluted rivers the oxygen content may fall to zero. At Walvis Bay on the west coast of South Africa low or zero oxygen contents exist over several hundred square miles of ocean.

CARBON DIOXIDE

The amount of free carbon dioxide in natural waters is seldom greater than 10 ppm and a part of this is closely connected with the carbonate equilibria and will be discussed later. In condensate from boilers which are fed with water containing appreciable amounts of carbonate the concentration of dissolved carbon dioxide may be very high owing to decomposition in the boiler.

HYDROGEN SULPHIDE

Amounts of hydrogen sulphide of up to 15 ppm may occasionally be present owing to pollution or, more often, the action of sulphate-reducing bacteria. As little as 0.5 per cent may be detected by its objectionable odour.

Both carbon dioxide and hydrogen sulphide may be reduced to an acceptable figure by aeration.

ORGANIC MATTER

The organic matter present in water consists of living organisms and the products of their metabolism or decay.

NON-LIVING ORGANIC CONSTITUENTS

These may be in colloidal or true solution or in suspension and are derived from decay of vegetable matter in the drainage area, domestic and industrial wastes, and oil contamination.

In water from moor and marsh land, besides carbonic and humic acids, citric, acetic and benzoic acids are found. These extracts from peat and mosses lower the pH value of soft waters and produce discolouration. They increase the corrosive action on ferrous metals as well as on lead and copper.

LIVING ORGANIC MATTER

There are two main classes of living organisms found in waters; microscopic organisms such as bacteria, slimes, fungi and algae, and macroscopic marine organisms such as barnacles. In their metabolism such organisms produce significant changes in the composition of the water in their immediate neighbourhood. Thus algae remove carbon dioxide and give off oxygen while other organisms consume oxygen. Hydrogen sulphide may be produced by sulphate-reducing bacteria or by the decay of organic matter. This decay may also give corrosive amino acids, e.g. cystine, di-αamino-β thiopropionic acid, which deposit deleterious sulphide films on copper-alloy condenser tubes (see Chap. VIII).

Iron bacteria, on the other hand, do not change the composition of the water nor do they take part in the corrosion reaction. They do however produce fouling owing to their accumulation of large amounts of ferric hydrate which may be up to 500 times as great as the volume of bacteria concerned. This leads to blockage, increase in friction owing to the formation of tubercles, and the production of "red water". Iron bacteria may be controlled by adding sodium carbonate to give a pH of 8.5 or by chlorination to give a residual chlorine content of 0.5 ppm.

Marine organisms such as barnacles and molluscs have a marked fouling action and attach themselves to metal surfaces providing the water is stagnant or slow moving. The firmness of attachment is greatest on smooth hard surfaces and least on poorly adherent corrosion products. Such organisms can cause damage to protective coatings of paint or bitumen due to their secretions or to mechanical effects on the coatings. Certain barnacles can damage bituminous enamels 250 mils thick and, in waters where such organisms proliferate, anti-fouling paint should be applied at regular intervals.

As with any type of deposit the influence of fouling is dependent on the completeness of the coating. If the coverage is complete reasonably good protection will be afforded. Such deposits will, however, create the necessary anaerobic conditions for sulphate-reducing bacteria to become active provided the necessary sulphate and organic nutrients are available and the pH value is between 5.5 and 8.5. If the coverage is only partial, oxygen concentration cells can arise. Fouling by living organisms can be prevented if the copper dissolution from the metal or anti-fouling paint is not less than 5 mdd under relatively quiescent conditions. Algae may be restrained by shutting off light or killed by chlorination or addition of copper sulphate.

In boilers the deposits mentioned can lead to over-heating which may eventually result in the bursting of tubes. Similarly a very small amount of oil on a heating surface can lead to very rapid failure.

BICARBONATE EQUILIBRIA

Carbon dioxide dissolves freely in water with a solubility depending on the partial pressure, i.e. its concentration in the gas phase. In air the partial pressure of carbon dioxide is 0.0003 atm. and the concentration of carbon dioxide in water in equilibrium with air is about 0.5 ppm. On the other hand if the gas phase above the water is wholly carbon dioxide the amount dissolved is over 1,500 ppm.

Most of the dissolved carbon dioxide is present in solution in the molecular state. A small amount reacts with the water to form carbonic acid,

$$CO_2 + H_2O \rightleftharpoons H_2CO_3$$

with $\dfrac{[H_2CO_3]}{[CO_2]} = 0.003$

The carbonic acid dissociates in two stages, first to form bicarbonate ions, HCO_3^-, and subsequently carbonate ions, CO_3^{--}, according to the equations,

$$H_2CO_3 \rightleftharpoons H^+ + HCO_3^-$$
$$HCO_3^- \rightleftharpoons H^+ + CO_3^{--}$$

The first and second dissociation constants corresponding to the above equations are given by

$$\frac{[H^+][HCO_3^-]}{[H_2CO_3] + [CO_2]} = K_1 \tag{1}$$

$$\frac{[H^+][CO_3^{--}]}{[HCO_3^-]} = K_2 \tag{2}$$

At 25°C, $K_1 = 4.45 \times 10^{-7}$ and $K_2 = 4.69 \times 10^{-11}$

From this data the variation in the ratios of the various components may be calculated from the equations

$$\log \frac{[HCO_3^-]}{[H_2CO_3] + [CO_2]} = pH - 6.35$$

and $\log \dfrac{[CO_3^-]}{[HCO_3^-]} = pH - 10.33$

for any given value of the pH. The variation in the relative percentage of the total dissolved carbon dioxide in the various forms is given in Fig. 14. It will be seen that up to a pH value of about 8.5 we have only to consider the equilibrium between "free carbon dioxide" and bicarbonate, the amount

of carbonate being negligibly small. In more alkaline solutions we only
need to consider the ratio of carbonate to bicarbonate. In pure water in
equilibrium with the air the pH value will be 5.7 while for water in
equilibrium with an atmosphere of carbon dioxide the pH value will be
about 3.8.

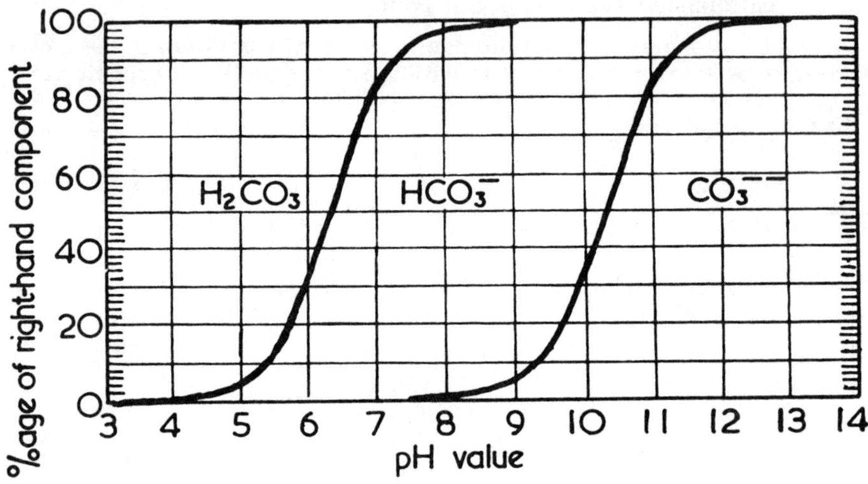

Fig. 14.   BICARBONATE EQUILIBRIUM AND pH

Natural waters have pH values in the range 6.0 to 9.5 but 90 per cent of
the waters used have values between 6.5 and 8.0 (see Fig. 12). Thus we
are generally concerned with the free carbon dioxide—bicarbonate equilibria.

ALKALINITY AND HARDNESS

The amount of hydroxide, carbonate and bicarbonate present in the water
is referred to as the alkalinity (Alky)

i.e. $\text{Alky} = [H_2CO_3] + [HCO_3^-] + [CO_3^{--}] + [OH^-]$ \hfill (3)

This is usually estimated in two stages by titration, first with phenolphthalein as indicator followed by titration with methyl orange. As will be
seen from Fig. 14 the alkalinity to phenolphthalein, which gives an end-point
at 8.0, is a measure of the hydroxide and half the carbonate alkalinity.
The alkalinity to methyl orange, which gives an end-point at 4.5, corresponds to the carbonate, bicarbonate and hydroxide, or total alkalinity. A
knowledge of these two values allows the amounts of bicarbonate and carbonate to be calculated.

Calcium and magnesium are the constituents producing <u>hardness</u> in the water and are able to destroy the detergent properties of soaps. The calcium and magnesium salts of the long-chain fatty acids in the soap are very insoluble and destroy its lathering properties by producing a scum. These metals are generally present in solution as the bicarbonate but they may also occur as the sulphate or nitrate and occasionally as the chloride. Their presence is very important in steam-raising or hot water installations owing to their ability to form scales.

The <u>total hardness</u> is best determined by direct estimation of the calcium and magnesium contents. Formerly the term <u>temporary hardness</u> was applied to that part of the hardness precipitated on boiling, i.e. associated with the carbonate and bicarbonates of calcium and magnesium. The figure obtained is, however, dependent on the technique used and is being replaced by the term <u>carbonate hardness</u> which is generally the same as the total alkalinity. The difference between the carbonate hardness and the total hardness, estimated from the calcium and magnesium, is called <u>non-carbonate</u> and often replaces the older term <u>permanent</u>. This part of hardness is produced by the sulphates, chlorides and nitrates of calcium and magnesium.

Magnesium is generally a minor constituent compared with calcium but it is present in waters of the southern part of England as well as the soft waters of the north. In certain underground waters drawn from older geological formations, in particular coal measures, the magnesium content may be high and equally important to calcium.

The main cations present, other than calcium and magnesium, are sodium and potassium derived from deposits of their carbonates, sulphates or chlorides. Potassium is not often present in large amounts and the sodium content is usually less than the calcium. Larger amounts of sodium can often occur in waters drawn from older geological formations. Waters drawn in the London Basin from chalk and sands beneath the clay are unusually rich in sodium salts, partly attributed to a natural base-exchange softening. Many brackish waters, and in particular sea water, contain amounts of sodium appreciably higher than the calcium or magnesium.

Although the hardness of British waters (expressed as calcium bicarbonate) may reach figures of the order of 600 ppm, 42 per cent are below 100 ppm and 26 per cent have a hardness in the range 400-500. The actual distribution of hardness and its relationship to geological formations is shown in Fig. 13.

THE LANGELIER INDEX AND SCALE FORMATION

The solubilities of the principal carbonates in water (expressed as ppm calcium carbonate) are: calcium carbonate, $CaCO_3$, 13; magnesium carbonate, $MgCO_3$, 75; sodium carbonate, Na_2CO_3, 289,000. The amount of carbonate present in water is controlled by the solubility of calcium carbonate, i.e. by the solubility product, K_S

$$Ca^{++} + CO_3^{--} \rightleftharpoons CaCO_3(\text{solid})$$

$$[Ca^{++}][CO_3^{--}] = K_S \tag{4}$$

From a combination of equations (1) to (4), together with the dissociation constant of water K_W,

$$[H^+][OH^-] = K_W$$

we obtain the following equation relating the concentration of the participating ions at equilibrium

$$\text{Alky} + [H^+] - \frac{K_W}{[H^+]} = \frac{K_S}{[Ca^{++}]} \cdot \frac{[H^+]}{K_2} \left\{ 1 + \frac{2K_S}{[Ca^{++}]} \right\}$$

Taking logarithms and rearranging we find that a solution saturated with calcium carbonate will have a pH value given by

$$pH_S = pK_2 - pK_S + pCa^{++} + p\left[\text{Alky} + [H^+] - \frac{K_W}{[H^+]} \right] - p\left[1 + \frac{2K_2}{[H^+]} \right]$$

where p represents the negative logarithm of the appropriate quantity. The values K_W, K_2 and K_S are known over a wide range of temperature. Hence if the calcium hardness, the total alkalinity, the pH and temperature of the water are known, the saturation pH, pH_S, may be calculated from this equation or determined from curves constructed from it. The difference between this and the actual pH, i.e. $pH - pH_S$, is known as the Langelier Index for the water at a given temperature. If this is positive, deposition of calcium carbonate from the water will occur while if it is negative the water will be capable of dissolving any calcium carbonate with which it comes into contact until it is saturated.

In the absence of adequate analytical data for the calculation of the Langelier Index the Marble or Chalk Test may be used. In this the water is equilibrated with chalk or marble and the change in pH and alkalinity will show whether calcium carbonate has been taken into solution or deposited.

From Fig. 14 it will be clear that in most waters the calcium carbonate will be in the form of the bicarbonate. With a few exceptions, where the amount of magnesium or sodium is unusually high, calcium bicarbonate is responsible for the <u>alkalinity</u> of natural waters and also for their <u>carbonate</u> or <u>permanent</u> hardness.

The solubility of calcium carbonate in natural waters is very dependent on the amount of dissolved carbon dioxide as is clear from the equation showing the first dissociation of carbonic acid and the data shown in Table 3.

TABLE 3

DEPENDENCE OF THE SOLUBILITY OF CALCIUM CARBONATE
ON THE PARTIAL PRESSURE OF CARBON DIOXIDE AT 25°C

CO_2 partial pressure, atmospheres.	$CaCO_3$ solubility, ppm.
0. 0003, as in air.	53
0. 001	78
0. 01	170
0. 1	390
1. 0	900
10. 0	2250

The amount of calcium carbonate dissolved by water percolating through
the rock and soil will, therefore, be a function of the concentration of
carbon dioxide derived either from the atmosphere or from the decomposition of organic matter. The equilibrium between the solid calcium
carbonate and carbon dioxide may be represented by the equation,

$$H_2O + CO_2 + CaCO_3 \text{ (solid)} \rightleftharpoons Ca(HCO_3)_2$$

This is a reversible reaction and deposition of calcium carbonate will
occur if the free carbon dioxide content of the water is reduced. Indeed,
calcium bicarbonate is only stable if a certain concentration of carbon dioxide is maintained.

The amount of calcium bicarbonate in natural waters varies from very
small amounts in waters from upland sources to the 200 to 300 ppm (as
calcium carbonate) in those obtained from underground or lowland sources.
In exceptional cases the amount present may be more than 300 ppm and
a few underground supplies exist in which it is as high as 350 to 400 ppm.
These concentrations are associated with a proportional amount of carbon dioxide and the pH value as shown in Fig. 14.

Rain water which, in unpolluted atmospheres, is the purest natural water
usually has up to 0. 5 ppm carbon dioxide and a low pH value. The amount
of dissolved carbon dioxide can be increased still further by contact with
decaying carbonaceous matter. Often the amount of carbon dioxide is
greater than that necessary to stabilise the calcium carbonate. Normally
waters have a bicarbonate alkalinity and the ratio of the free carbon
dioxide to the methyl orange alkalinity is shown in the pH value:-

$$\log \frac{M.O. \text{ alkalinity}}{\text{Free } CO_2} + 6. 35 = pH$$

In other words, when the ratio M.O. alky/Free CO_2 is 1.0 the pH is 6.35, when 10.0 the pH is 7.35 and when 100.0 the pH will be 8.35.

If air is bubbled through waters of increasing methyl orange alkalinity an increasing percentage is converted to phenolphthalein alkalinity until the latter reaches a stable figure of 7 per cent of the M.O. alkalinity. When the M.O. alkalinity is 100 ppm there is 86 ppm bicarbonate alkalinity and 14 ppm carbonate alkalinity (twice the phenolphthalein alky).

The deposition of scale is a very important property of a water from a corrosion point of view since such scales can have a marked restraining action on the progress of corrosion. If this deposition is excessive it can result in appreciable reduction in pipe diameter and in an extreme

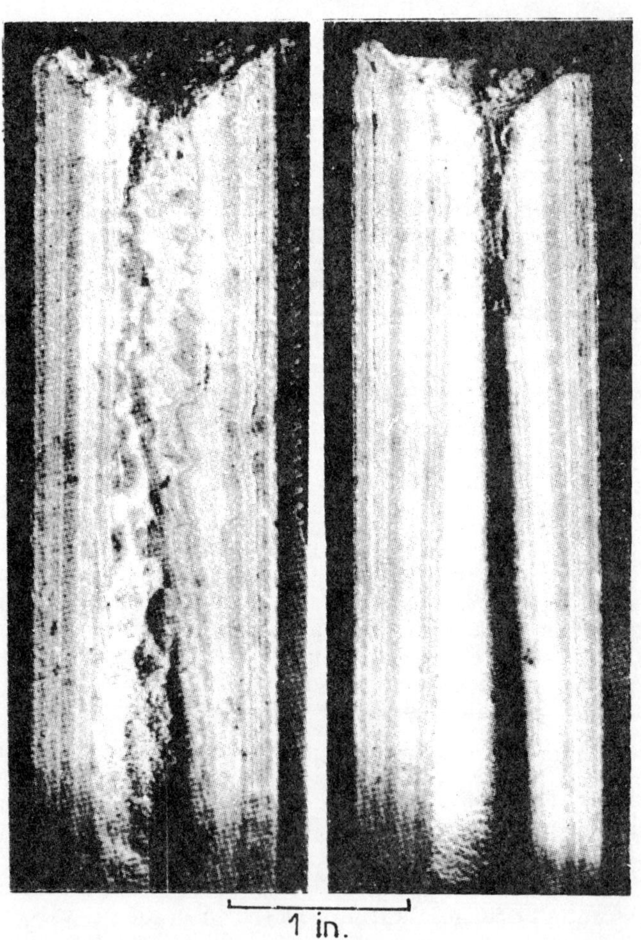

Plate 1.   SCALING OF BOILER OUTLET IN STEAM BAKERY.

case can completely stop the flow of water (Pl. I). The calcium carbonate
in association with the products of corrosion can often produce a layer
which severely restricts and even completely prevents attack. It is, therefore, fortunate that many waters used industrially are supersaturated in
calcium carbonate, owing to their high content in carbon dioxide, but
when they come into contact with aerated surfaces the calcium carbonate
is precipitated. Furthermore, on the surface of a corroding metal the
alkali generated by the cathodic reduction of oxygen,

$$O_2 + 2H_2O + 4e \rightarrow 4OH^-$$

raises the pH on part of the surface which converts the bicarbonate to
carbonate with a consequent deposition of carbonate scale. Such deposition
can occur even if the solution is not supersaturated in calcium carbonate.
The physical nature of the carbonate deposited may also be influenced
by very small amounts of organic matter present in the water because
waters with identical inorganic salt concentration can give entirely different deposits of calcium carbonate; a friable deposit with little or no
protective value or a scale of egg-shell thickness which virtually prevents all further attack.

The solubility of many salts decreases with increase in temperature.
Important among these are calcium carbonate and calcium sulphate.
The effect of increasing temperature on the solubility of calcium carbo-
nate is shown in Table 4.

TABLE 4

EFFECT OF TEMPERATURE ON THE SOLUBILITY OF
CALCIUM CARBONATE

Temp., t°C.	Ratio of solubility at t°C. to that at 25°C.	Calculated solubility of $CaCO_3$, ppm, at CO_2 pressure of.	
		0.0003 atm.	1.0 atm.
0	1.8	95	1600
10	1.4	75	1250
20	1.1	59	1000
25	1.0	53	900
30	0.9	47	800
50	0.6	32	550

It is readily apparent that a relatively small change in temperature will
cause calcium carbonate to be precipitated. The solubility of calcium

sulphate at ambient temperature is over 1, 300 ppm and the solubility increases with increase in temperature up to 40°C. On further increase in temperature however, the solubility decreases rapidly and at 200°C is only 56 ppm. In boilers, precipitated calcium sulphate forms very hard scales which are very difficult to remove.

WATER ANALYSES

These can be extremely confusing as it seems that the chemist has tried to present them in as many different ways as possible. To be fair, one must admit that the concern of the water chemist is not that of corrosion, *per se,* but of hardness and, above all, potability. Not satisfied with juggling with the units alone which may be ppm, pp 100, 000 or grains per gallon (and in the U.S A., U.S. grains per U.S. gallon!) he presents the anions and cations separately or joined together into a suggested composition of salts dissolved in the water, thereby creating an endless variety of combinations. It is common now to express all the anions, cations, and alkalinities in terms of $CaCO_3$ which has the convenient molecular weight of 100 making calculations easier. Even so, from the previous parts of this chapter it will be appreciated that to those interested in corrosion there is something to be said for having the Cl^-, SO_4^{--}, Ca^{++}, TDS and free CO_2 quoted simply as ppm with the alkalinity and hardness as ppm $CaCO_3$.

Apart from the difficulty of units, it is as well to point out that the site and time for collection of water samples must be chosen with care so as to reflect the corrosive conditions, e.g. a river or canal subject to tidal effects needs a yearly survey of the variations in composition.

INDUSTRIAL WATERS

NATURAL WATERS

Rain water is the purest of all natural water supplies. In rural areas the only dissolved constituent other than oxygen is carbon dioxide which renders it slightly acid and more aggressive. The aggressivity is increased in industrial areas by absorption of sulphur oxides which react with water to form a dilute solution of sulphuric acid. The low pH of such waters makes them aggressive to calcareous matter and not only can they cause deterioration of stone and mortar but also, in common with all soft waters, they can dissolve protective carbonate scales from a metal surface.

Although the original water is very low in total dissolved solids these may be considerably increased during collection and storage. Dissolution of solids from roofs and rain water gullies, from concrete tanks and of air-borne solids deposited during storage all increase the amount of dissolved material and the hardness of the water.

Upland surface waters are similar in composition to rain water. Where the watersheds are of igneous rock the amount of mineral matter picked up by the water is small. Such waters are mainly confined to the north and north-west of the British Isles. A typical water may have a pH of 6. 0 or less and a total hardness up to 20 ppm, 0 to 50 per cent of which may be attributed to carbonate hardness. Such waters are aggressive to most of the common metals.

In considering surface waters such as rivers and streams the composition will depend on the nature of the land through which they flow. This also determines the composition of the reservoirs which they may feed. The land may be pasture, arable, peat, almost insoluble rock, all types of limestone and any type of sand. Furthermore the water may be polluted by sewage or any trade waste or effluent. It follows that these waters can vary greatly in composition. As a general rule, if the drainage area is virtually insoluble the water will be soft; if there is considerable de- caying vegetation or peat the water will be acid; and if the water passes through chalk it is likely to be hard. In Scotland the Dee draining through largely insoluble granite is soft, while the Thames receiving its water from chalk is hard and the Trent, fed by streams passing through magne- sium-bearing limestone, contains relatively large amounts of magnesium salts.

The surface waters are maintained by rain water which, as we have al- ready mentioned can contain varying amounts of carbon dioxide derived from the air but in larger amounts from decaying organic matter. From what has already been said it will be clear that the solution by water of minerals such as limestone, chalk, dolomite and magnesite is very de- pendent on the carbon dioxide content. Calcium carbonate has a very limited solubility in water but dissolves readily as the bicarbonate in carbonated waters. Sulphates, chlorides and nitrates are readily soluble in water.

Tidal rivers vary in composition over a very wide range and the chloride content will not only rise and fall with the tide but will also change with the season. In the rainy seasons the tide is pushed back by the flood water but in the dry season the tidal inflow is greater with a correspond- ing increase in chloride content. Thus the chloride content would be ex- pected to rise from, say, March to September, and fall during the autumn and winter months.

Underground supplies from wells and springs are of greatest importance in the south and east of England. Wells are broadly classified into "shallow" and "deep" depending on the absence or presence of an im- pervious layer above the water. Shallow supplies are most vulnerable to pollution, e.g. in fissured chalk with swallow holes the underground water has similar characteristics to the surface supply. Waters from deep wells, which are usually greater than 100 feet in depth, have a high stan- dard of bacterial and organic purity. They are usually free from oxygen and hence may carry appreciable iron in solution. The composition and temperature of well waters are more reproducible than for other supplies.

If the strata are such that sea water can seep into the wells— and this is
quite likely in coastal regions—the wells may contain very high concentrations of chloride. An artesian well in New Windsor contained 320 ppm
chloride (TDS 1, 250 ppm), a borehole water at Harwich 1, 132 ppm (TDS
2, 577 ppm) while a well water at Awali in Bahrein contained no less than
7, 310 ppm (TDS 15, 000). Many wells contain large amounts of free carbon dioxide and, like the high chloride waters, may be very corrosive.

Mine waters are not a widely used source of supply but their quality is
obviously important in this particular industry. Some of them are very
little different from surface waters. They may vary widely according
to the strata through which the water drains and may even vary in composition in the same mine. A number of them are very acid and, as a result very corrosive.

Sea water, with total dissolved solids of about 35, 000 ppm, of which about
20, 000 ppm are chloride ion, is the natural water with the highest concentration of dissolved salts. Owing to its availability, abundance and
cheapness it is widely used at sea and in coastal regions for cooling purposes and, owing to the large interest in the corrosion of sea-going vessels its corrosivity has been the subject of much investigation.

TREATED WATERS

Purified Water.   Boiler condensate and distillate and also deionized
waters are of similar or higher purity than rain water. The condensate
or distillate may have a much higher content of dissolved carbon dioxide
if the feed water contains appreciable amount of carbonate or is treated
with sodium carbonate, which decomposes at high temperatures to give
carbon dioxide. This makes the water particularly aggressive to metals
such as copper and iron. In the process of deionization carbonates are removed and the resulting water is low in carbon dioxide and its aggressivity will be largely determined by the amount of dissolved oxygen.

Softened Waters.   In many industrial processes it is necessary to soften
or purify the water. One of the cheapest ways of softening is by addition
of lime and sodium carbonate. The calcium and magnesium are precipitated as calcium carbonate or magnesium hydroxide but the sulphates and
chloride remain in solution as the sodium salts.

Zeolite softening is carried out by passing the water through a bed of
sodium alumino-silicates. The calcium and magnesium ions exchange
with the sodium ions in the zeolite and the scale-forming salts are converted into the very soluble sodium salts. When exhausted the zeolite may
be reconverted to the sodium form by use of a solution of sodium chloride.

The use of zeolites for softening is gradually being replaced by the use
of synthetic resins which are able to exchange ions. In the cation exchange
resins all the cations, viz. calcium, magnesium and sodium, are converted
into hydrogen ions. The carbonic acid formed from the carbonates may
be completely removed as carbon dioxide but the acids formed from the

chloride and sulphate must either be neutralised or removed by an anion exchange resin. The cation exchange resin is reactivated with hydrochloric acid.

The <u>anion exchange</u> resins replace the anions in the water with hydroxyl ions. They may be reactivated with caustic soda.

By the use of a mixed bed of these two resins or by passing the water through them in series all the dissolved salts are removed at the same time giving a very pure water of low conductivity. Organic materials are not, however, removed.

Composition Changes during Use. In use, the composition of the water often changes either by evaporation from cooling towers and spray ponds or by extraction of steam in steam-raising equipment. This will result in an increase in dissolved solids and, unless water is regularly bled off, the concentration may become very large indeed. Furthermore the content in aggressive salts such as chloride and sulphate increases continuously owing to their high solubility while scale-forming salts, having a low solubility, are precipitated. This results in an increase in aggressive ion concentration and a decrease in the concentration of potentially restricting ions. The amount of water which can be bled off represents a balance between losses due to corrosion and the cost of replacement water. Concentration factors of between five and ten may be expected and this should be taken into consideration in assessing the probable corrosive properties of a given water.

Chapter III

FORMS OF CORROSION

The form, appearance and distribution of the attack, in particular the degree of localisation, and the topography of the affected area are all important characteristics of different types of corrosion.

UNIFORM ATTACK

Reaction of the metal with water to give a uniform depth of penetration of the metal over the whole surface is the most satisfactory state of affairs with which the engineer has to deal since due account can be made at the design stage for the thinning and decrease in cross-section. This type of attack is commonly encountered in acid solution and on some metals in strongly alkaline solution. It is more likely to be encountered in waters with a high content of dissolved salts, i.e. of high electrical conductivity, in which the action of local anodes and cathodes is not so readily encouraged. A surface covered with a uniform deposit is more likely to have a reasonably uniform type of attack. If the deposit is calcium carbonate, as would occur in hard waters, the rate of attack will also be reduced to a small or negligible value.

GROOVING

Appreciable thinning and grooving of the metal, which is almost free from corrosion product, is characteristic of attack by acid condensate, i.e. condensed steam high in dissolved carbon dioxide. In condensers, the grooves originate from the areas where the steam first starts to condense and are indicative of the flow path of the condensate. This form of attack is found on steel and copper condenser tubes, (Pl. 2), and can be caused by large amounts of carbonate in the feed water to the boiler or in the boiler water treatment. Severe thinning of the lower half of horizontal returns is caused by the flow of condensate through partly filled pipes or condensate remaining in the pipes after shut-down.

Grooving can also occur at a bimetal junction in a thin layer of water or in water of low conductivity when attack will be largely concentrated on the anodic metal close to the line of contact. When deeply immersed in highly conducting media the effect of the couple is spread over greater distances and the localisation of attack is less evident.

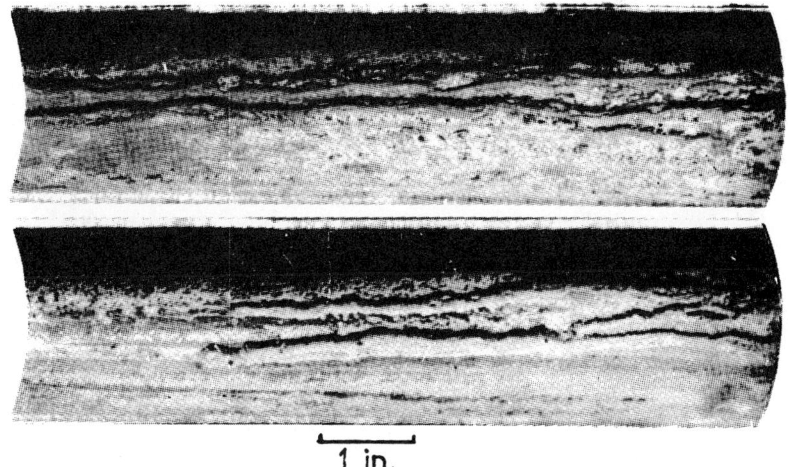

Plate 2.   GROOVING OF COPPER CALORIFIER TUBES BY
ACID CONDENSATE. 15 YEARS.

PITTING

Pitting is a dangerous form of localised attack in which the depth of penetration is of at least the same order as the diameter of the area corroded.
This type of attack can often result in the perforation of a pipe or vessel
even though the amount of metal lost due to corrosion is very small.

The pits may vary from shallow hemispherical forms to pin-hole pitting
in which the depth of penetration is far greater than the diameter.

The intensity of localisation may be expressed in terms of the pitting
factor which is the ratio of the greatest depth of penetration to the average
depth of penetration calculated from the weight loss. A pitting factor of
unity is indicative that the corrosion is uniform. Unfortunately in most
cases localised attack occurs and this factor becomes quite large. The
use of this factor is only justifiable where the attack is confined to pitting.
Roughening and general attack often accompany pitting and such a factor
can then be positively misleading. In Fig. 15 the first example has a pitting factor of 1.82 while in the second the factor is 1.29 although the
depth of penetration is twice as great.

It is often necessary to express the localisation of attack simply in terms
of the deepest pit. This has a real basis in practice since if a tube or tank
is holed there is little merit in knowing either the distribution of pitting or
the average of an arbitrary number of the deepest pits. There is much
information on the overall corrosion rates of metals in a variety of media
but data on the pitting factor and the rate of penetration of pits is very
sparse indeed.

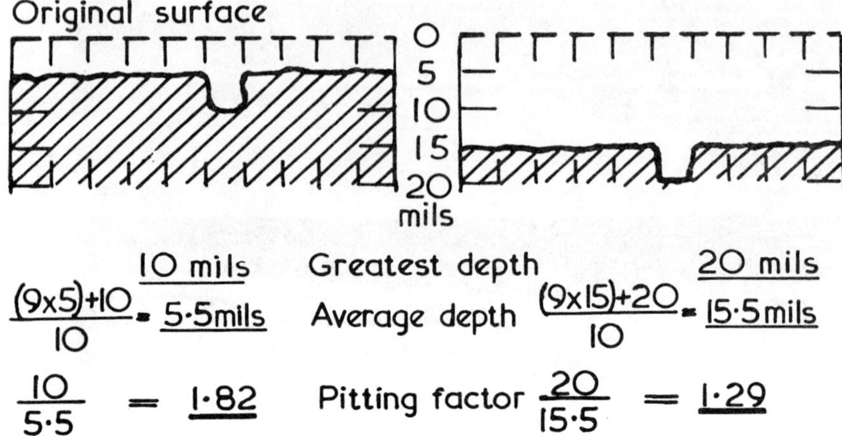

Fig. 15.   VARIATION OF PITTING FACTOR WITH AMOUNT OF ATTACK

Although the phenomenon of pitting is imperfectly understood it is associated at its inception with a small anodic area and a large cathodic area. The rate of attack at the small anode is a function of the conductivity of the solution and the cathodic and anodic polarisation. In a poorly conducting water the effective cathode will be the area of metal in the immediate neighbourhood of the pit, but in a highly conducting medium such as sea water the cathodic reaction may be occurring at a considerable distance away (Chap. I).

The heterogeneity giving rise to pitting attack may exist in variations in the metal, in a film or layer at the metal surface and in the film solution interface. Metallic variations will be considered in Chap. VII. The cathodic film is often the oxide of the metal or a deposit of more noble metal or carbon. Any local breakdown of such a film will initiate intense local attack and pitting. The film may be laid down in manufacture, e.g. the mill-scale on iron and steel, a common cause of intense localised attack on these materials. (Plates 3-6).

Considerable trouble was experienced some years ago with copper cold-water pipes in hard waters. This was eventually attributed to a film of carbon which had been derived from the lubricant used in the drawing process and which acted as a very efficient cathode. In other cases local breakdown of thin glassy films of cuprous oxide produced during heating, e.g. in brazing, are responsible for pitting. Pitting of the copper sheathing of a yacht was ascribed to the formation of a visible red copper oxide film produced by the use of an unsuitable metal primer in painting. Following paint failure, rapid penetration of the sheathing occurred. Troubles with the pitting of copper in hot-water systems using soft manganese-bearing

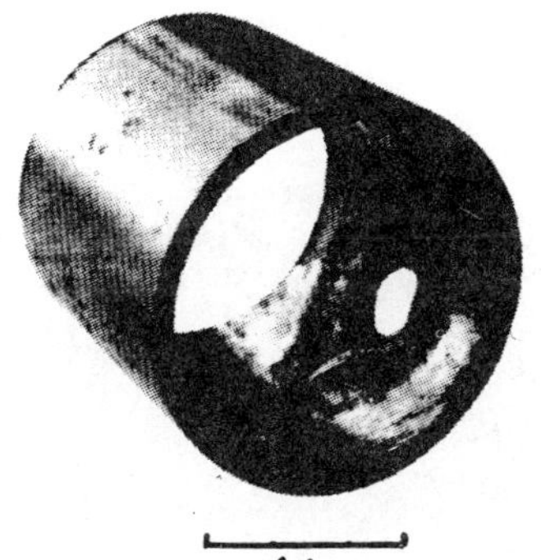

Plate 3. MILD STEEL CONDENSER TUBE, LOCALISED ATTACK DUE TO MILL SCALE, IN BRACKISH ESTUARY WATER. 6 MONTHS.

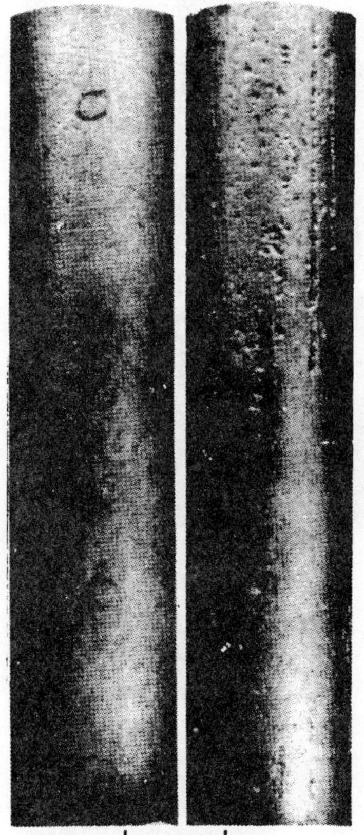

Plate 4. TESTS SHOWING THE INFLUENCE OF MILL SCALE. TWO WEEKS IN BRACKISH WATER, TUBES ROTATED AT 2,000 rpm. DELIBERATE PUNCH MARKS ON ORIGINAL TUBE SHOWN RINGED. TUBES CLEANED AFTER TEST.

Plate 5.   LOCALISED ATTACK ON BORE-HOLE PIPING
OWING TO MILL SCALE. HIGH CHLORIDE WATER.
2 YEARS.

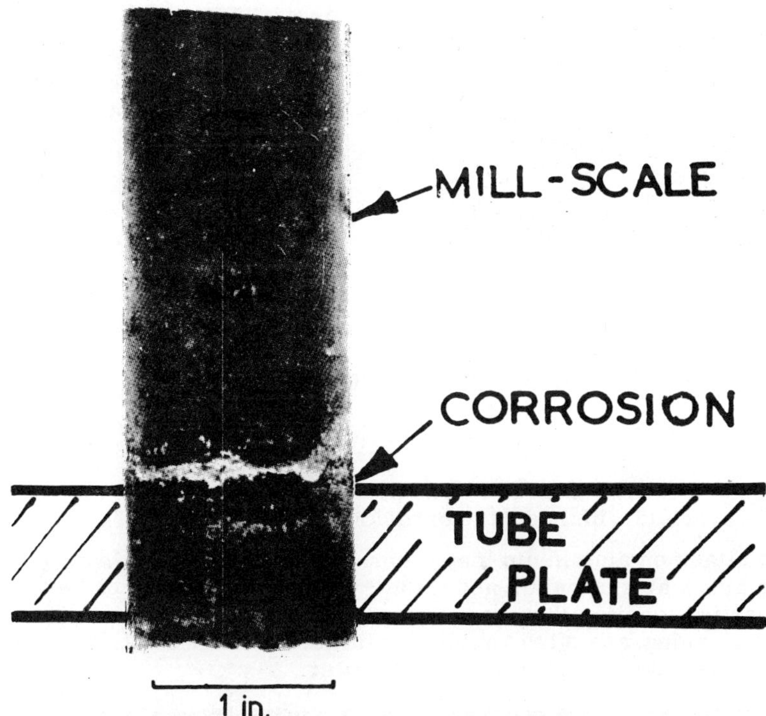

Plate 6.   LOCALISED ATTACK BY ACID CONDENSATE ON
CALANDRIA TUBE WHERE MILL SCALE HAS
BEEN REMOVED IN EXPANDING.

water was attributed to the hard scale rich in manganese oxides which
had been laid down over a period of several years and acted as an efficient cathode.

Copper-bearing waters can often be the cause of localised attack on a
number of metals. The critical amount of copper, necessary in the
natural water or dissolved from copper or copper-alloy parts in the
system, decreases in the order iron, zinc, and aluminium, the latter being
sensitive to a concentration as low as 0.02 ppm. Provided there is also
dissolved oxygen, calcium bicarbonate, and chloride present—which is
usually the case—this and larger amounts of copper will induce localised
pitting on aluminium (Plate 7). It may usually be recognised by the existence of a bluish tinge or flecks in the white corrosion product.

It is relatively easy to check for copper if it is suspected as the pit
initiator. We may cite as an example a galvanised cooling-water tank on

Plate 7.   LAVATORY FLUSH PIPE, ALUMINIUM. LOCALISED
ATTACK PRODUCED BY COPPER IN THE WATER.

which intense localised attack had resulted in deep pits containing copper
compounds overlaid with nodular corrosion product (Chap XIII).

Localised attack on aluminium may be prevented by regular cleaning
which explains why aluminium and its alloys can be used successfully for
most domestic utensils. However, electric kettles, which are not amenable
to regular cleaning, are often rapidly perforated in houses with a copper
plumbing system and using cuprosolvent water.

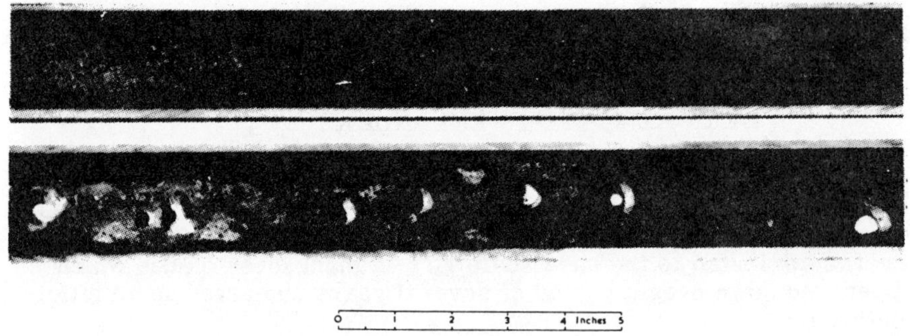

Plate 8.   LOCALISED ATTACK ON MILD STEEL TUBE IN
CLOSED HEATING SYSTEM. ATTACK IN LOWER
HALF ONLY, CAUSED BY SULPHATE-REDUCING
BACTERIA.

Attack by sulphate-reducing bacteria is readily recognised. The pits are
filled with a black deposit of sulphide which is easily washed away leaving
the metal clean. These pits are often hemispherical (Pl. 8) but may also
run into one another causing long grooves. When the system is freshly
opened there is also the distinctive smell of hydrogen sulphide but on ex-

posure to the air this is rapidly oxidised to sulphate and this evidence may
be lost. If confirmation is required it is preferable if the microbiologist
can be supplied with a fresh sample of product and water in a clean stop-
pered bottle.

Elongated pits are often encountered on stainless steel. The products
within a pit are salts, e.g. chlorides or iron, nickel and chromium, and the
pH there may fall as low as 1.0. This breaks down the passivity and any
flow either by gravity or otherwise causes elongation.

Pitting of stainless steel can be caused if waters become contaminated
with oxidising salts. The criterion of pitting is the oxidation-reduction
potential and if this is lower than -0.15 volts pitting conditions are created. This means that a slight change in operating conditions can cause
pitting when no attack may be expected.

The subject of pitting at high temperatures is discussed in Chapter VIII.

TUBERCULATION

In certain cases product from the corrosion of iron or steel pipes forms
mounds over the metal surface. (Fig. 7. Pl. 9). These mounds are usually
closely associated with pitting and usually overlie anodic areas where
localised attack is occurring. The growths are often so large and numerous that the frictional resistance to flow is increased and the carrying
capacity of the pipe is considerably reduced. Efforts to maintain the rate
of flow mean higher pumping costs and, ultimately, cleaning of the pipes
is necessary followed by installation of water treatment, operations which
cost much in time and labour. This tuberculation often occurs in soft
waters with a high carbonic acid content, i.e. high bicarbonate alkalinity.

The tubercles usually comprise a hard outer crust of brown hydrated
ferric oxide, with possibly an admixture of calcium carbonate, and an
inner layer of black magnetite and occasionally green rust. This crust
physically separates the anodic area within the pit from the area outside
where oxygen reduction is taking place. Chlorides and sulphates concentrate within the pit where the solution is slightly acid irrespective of the
pH of the water outside. Within the tubercle the oxygen concentration is
much reduced and is likely to be zero. The acidity tends to accelerate
the rate of attack by providing a secondary cathodic reaction, i.e. hydrogen evolution. The hydrogen formed in this way will either dissolve in
the metal or diffuse out through the product into the main bulk of water.

Although at first such tubercles will have no effect on the rate of penetration of the metal, in time they often become less permeable to ions and
the rate of attack will then decrease.

The absence of oxygen beneath such tubercles will provide the necessary
anaerobic conditions for sulphate-reducing bacteria to proliferate. Thus,
even if attack with hydrogen evolution or oxygen reduction has been stopped, corrosion may continue in the presence of sulphate bacteria.

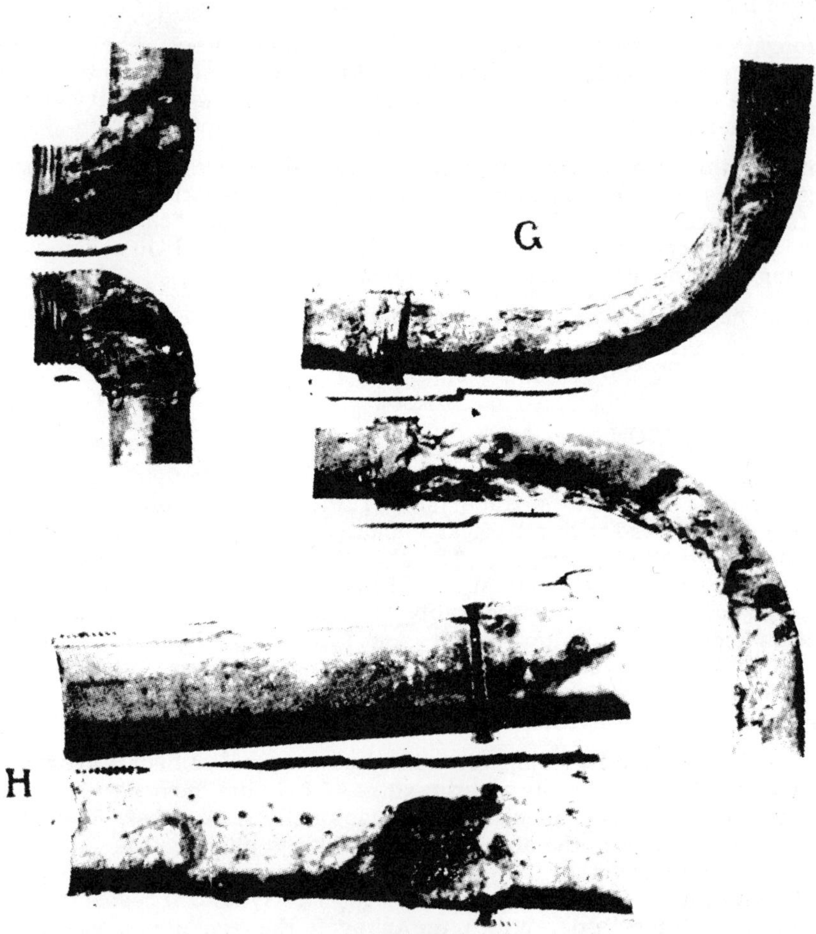

Before cleaning

Plate 9.   GALVANISED PIPES FROM HOUSING ESTATE
SHOWING INFLUENCE OF BENDS AND JUNCTIONS
ON CORROSION AND SCALE FORMATION

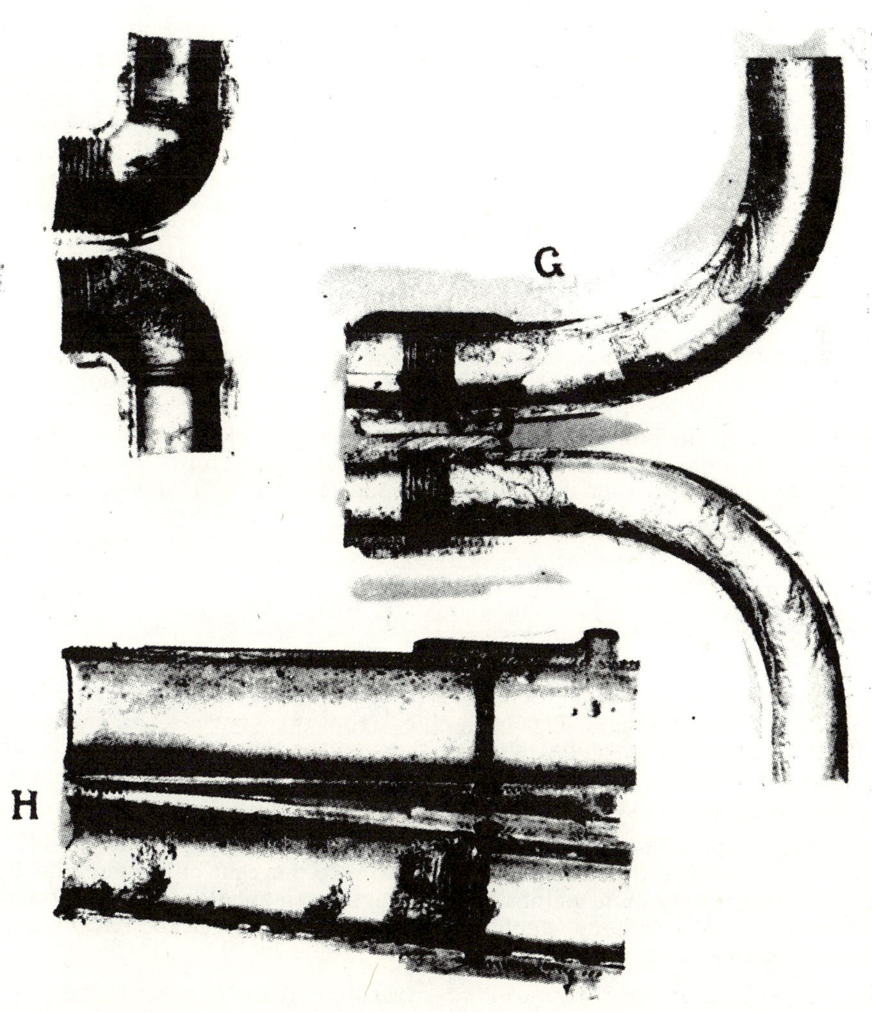

After cleaning

An unexpected form of attack may take place in flowing water containing sodium chloride and calcium bicarbonate. Long hollow whiskers of 4 to 6 mils in diameter are formed and grow to several inches in length in the direction of flow. These may be regarded as distorted tubercles in which the ferrous iron dissolved at the root of the whisker, forming a pit, diffuses along the tubelet and is oxidised to rust on coming into contact with the oxygenated solution at the growing tip.

Iron bacteria are a further cause of tuberculation and blockage of pipes. They do not take part in the main corrosion reaction but utilise ferrous iron in their metabolism and exuding ferric iron through their outer skin cause the accumulation of large quantities of hydrated ferric oxide. Such action may give troubles due to "red water".

EXFOLIATION

A peculiar form of attack can occur on 80/20 or 70/30 cupro-nickel tubes in high-pressure feed-water heaters. A severe form of general corrosion on the steam side consists of an exfoliating scale mainly of a mixture of metal oxides in the same proportions as the metals present in the alloy. The attack has only been observed on tubes heated intermittently and the mechanism is obscure. If continuous running is not practicable the only remedy is to use materials which are immune, i.e. copper or 90/10 cupro-nickel.

WATERLINE ATTACK

Localised attack can take place at the three-phase boundary, metal-water-air such as occurs at the waterline or at the periphery of a droplet of water on a metal surface. Trouble arises from two sources; the ready access of oxygen to the metal surface through the thin meniscus region and the difficult access of solution to the surface owing to the crevice formed between air/liquid surface and metal/liquid surface.

The greater access of oxygen to this region creates a concentration cell in which the area of the meniscus is cathodic to the rest of the surface. In the case of droplets on a metal surface there is an annular ring about 40 mils wide at the periphery which acts as a catchment area for oxygen and behaves as an efficient cathode.

As in the case of other forms of concentration cell the susceptibility to waterline attack varies with the metal and the solution. In chloride solution iron is not susceptible since the downward flow of cathodic/alkali will inhibit attack on the region immediately below the meniscus and rust will be precipitated at a lower level where the anodic and cathodic products meet. However, in hard waters the cathodic alkali causes the precipitation of calcium and magnesium salts. Such insoluble compounds will shield part of the iron from the aerated solution and induce local attack.

If an inhibitor such as carbonate or phosphate is added to the chloride solution the attack will become more and more restricted with increase in inhibitor concentration until ultimately the corrosion will be limited to a narrow band along the waterline. This effect is akin to that in crevice attack where access of inhibitor to the metal surface is restricted to diffusion over a relatively long path of small cross-section. Where a crevice or waterline is present, higher concentrations of inhibitor must be used.

In some natural waters the metal is protected at the waterline by a largely impermeable layer of oil absorbed on the surface, e.g. steel dams in polluted rivers. A further mitigating factor is the fluctuation in water level which has the effect of spreading out the attack over a wider area.

CREVICE ATTACK

Localised attack may occur in a crevice due to the limited access of electrolyte to the area within the crevice. This results in the formation of a concentration cell due to differences in salt, hydrogen ion or, more commonly, oxygen concentration.

Crevices can arise from contact of a metal with another piece of the same metal, other metals or any non-metallic material. Thus crevice attack can start in inaccessible corners produced by bad design and beneath foreign matter which settles on the surface. The latter may be carbonaceous matter, which can also act as cathode, but more commonly it is mud, sand, stone, cinders, iron scale, shells or cotton waste. Crevices can be formed by the intersection of wires, e.g. on a screen grid; when a tube is rolled into a tube plate e.g. in a calorifier or heat exchanger; within threaded junctions, e.g. screwed joints, and at an overlap, e.g. riveted seams (Pl. 10).

The shape of the crevice is important in determining the degree of accessibility of the electrolyte to the area within the crevice. It is important to know the angle of the crevice and whether the surfaces slope or curve away from one another. The intensity of attack within the crevice is dependent on the ratio of the area between the crevice and the exposed cathodic area. In a highly conducting medium such as sea water the extent and depth of attack on stainless steel is related to the area of surface outside the crevice.

The effect is greater with metals which may be passive or active such as stainless steel than with a readily corroding metal such as iron or steel. With iron, an anode at a point outside the crevice would develop more quickly than one inside the crevice owing to the greater proximity to the cathodic areas where oxygen reduction is occurring. In an inhibited solution, however, the behaviour of iron is similar to that of stainless steel in an uninhibited solution, i.e. intense attack takes place within the crevice. In the first case the cause is the lack of access of inhibitor to the affected area and in the second the insufficient supply of oxygen to keep the protective oxide film on the stainless steel in good repair. The molybdenum-bearing stainless steels are less susceptible.

Troubles from this cause rarely arise with aluminium presumably because when immersed in water this metal is able to form an oxide film even in the absence of oxygen. Titanium, another metal relying on an oxide film, is completely immune from crevice attack in sea water.

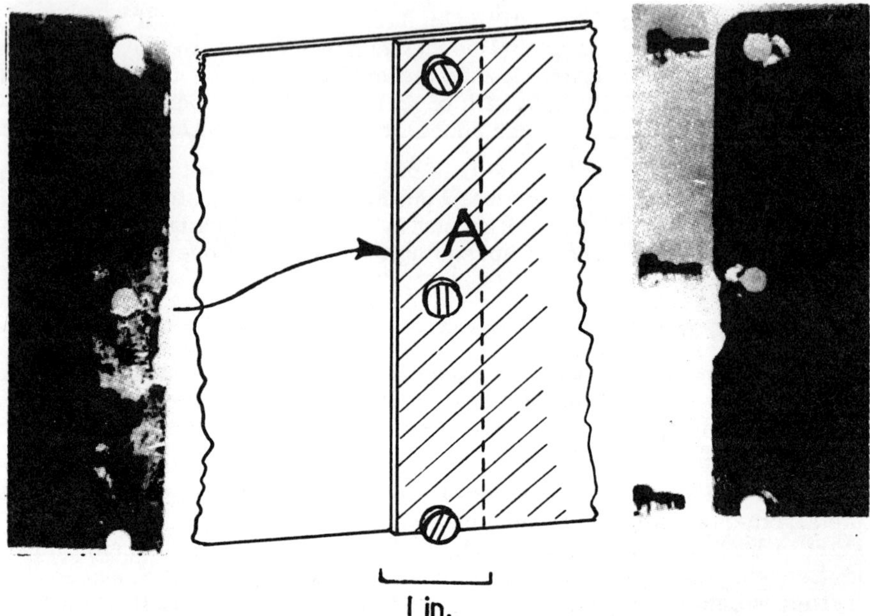

Plate 10. CREVICE ATTACK ON 18/8 STAINLESS STEEL IN
PURIFIED GELATINE AT 40°C. 6 WEEKS.

The conjoint action of galvanic effects and crevices should be guarded
against. Under normal conditions bronze and stainless steel would be compatible but in the case of a stainless steel axle running in a bronze bush
results are otherwise. The lack of oxygen supply to the stainless steel
reduces its potential and corrosion resistance to a similar figure to that
of iron and a powerful galvanic cell is then formed between the bronze and
the active steel. The similar bad influence of carbon particles on almost
any metal should not require any further clarification.

The actual behaviour of threaded and other joints depends on the tightness
of the joint. If this is tight and does not become wetted crevice attack will
not occur. The danger of this attack can be minimised by using a barrier
compound, even vaseline, in the joint or by the use of soldering or brazing,
all of which prevent liquid getting into the crevice.

Rubber gaskets can cause excessive attack at joints in stainless steel.
This can be overcome by combining metal powders with a rubber cement.
Zinc, tin, nickel, lead and ferrochrome prevent corrosion but magnesium,
aluminium, antimony, molybdenum, tungsten, copper etc. cause pitting.

Structures exposed to the atmosphere suffer more severe attack in
crevices since they remain wetted for a longer time, thus allowing corro-

sion to continue. The voluminous nature of the product is capable of levering plates apart and even breaking rivets. A similar condition might arise in a storage tank which is frequently left empty.

DEZINCIFICATION

The selective corrosion of brass, or <u>dezincification</u>, (see also Chap VII) is a common form of attack on brass in a number of waters, particularly sea water and soft water containing carbon dioxide. This results in the metal being converted into a porous mass of copper which is brittle and has poor mechanical strength. The only outward sign of this attack is a change in colour: instead of the yellow colour typical of brass the affected parts have the red colour of copper and are comparatively soft.

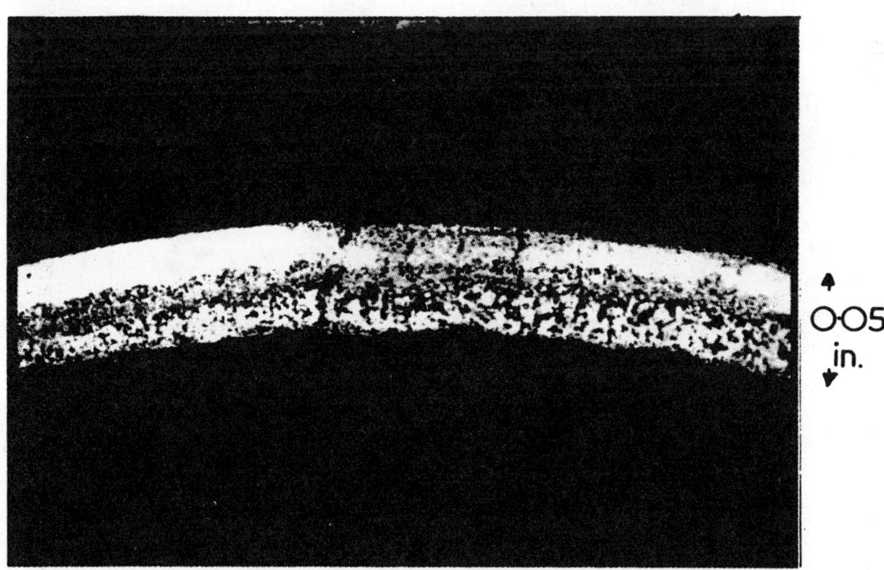

Plate 11.   "PLUG" AND "LAYER" TYPE DEZINCIFICATION
OF BRASS TOWEL RAIL IN SOFT, PEATY WATER.
11 years.

Two differing types of dezincification may be found, <u>plug</u> type, commonly found in sea water, and <u>layer</u> type, more often occurring in acid waters (Pl. 11, 12). In the plug type of attack the area affected is relatively small but the rate of penetration may be as high as 100 to 200 mils per year. Sometimes the attack may extend through the whole thickness of metal causing a leak, and occasionally the plug may be forced out leaving a hole. In the layer type of attack on the other hand the attack is uniform with a penetration rate of only 1 to 3 mils per year.

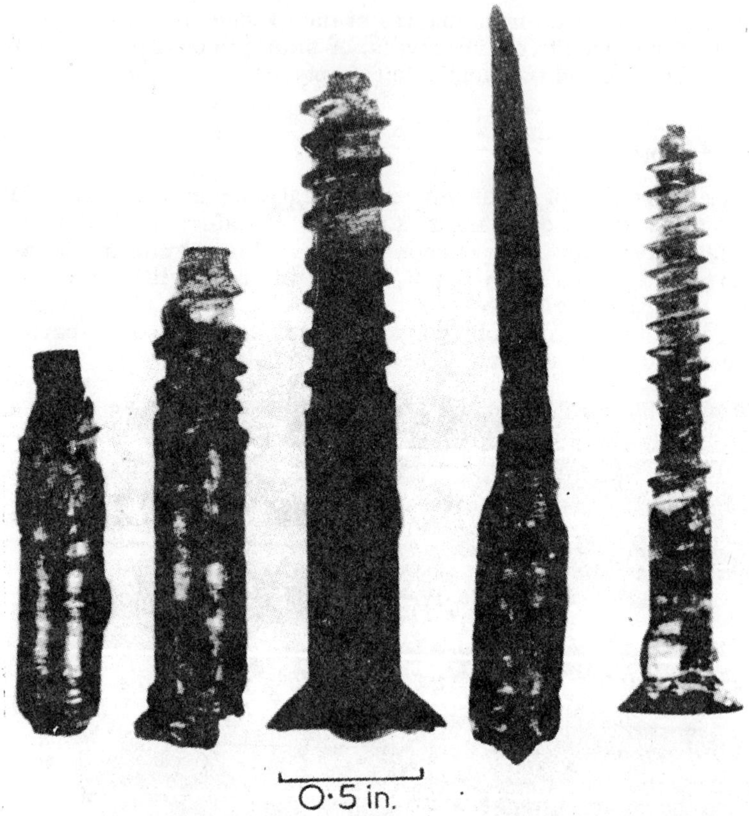

Plate 12. DEZINCIFICATION OF BRASS SCREWS BY SEA
WATER. LAYER TYPE.

The attack is facilitated by conditions of limited aeration such as exist
beneath deposits of corrosion product or sand and silt, in crevices and
where the flow rate of water is low. It is accelerated by high temperatures
and by high chloride concentrations such as occur in brackish waters and
sea water. In vertical sections of pipes the attack may be random but in
horizontal sections it will tend to occur at the bottom where debris and
deposits collect. Attack is often concentrated at screw fittings where
crevices exist and this leads to loss of strength and seizing of the joint.
This form of attack seldom occurs when the water velocity is high.

Similar forms of attack can take place with other copper alloys. Aluminium bronzes with 8 per cent or more of aluminium have a duplex structure and can suffer dealuminification in both acid and alkaline waters. The
aluminium-rich phase is preferentially attacked. Destannification of gun-

metal and phosphor bronze in superheated steam and of phosphor bronze pump impellers handling hot feed water and condensate have been encountered. The tin-rich phase of the duplex alloy is preferentially attacked. The reaction is slow on gunmetal but much more rapid on phosphor bronze. Decobaltification of cobalt alloys and denickelification of copper-nickel alloys can also occur under special circumstances but these are comparatively rare.

GRAPHITISATION

Both ductile and grey cast iron are susceptible to a form of attack known as graphitisation (see also Chap. VII) when immersed in waters which are slightly acidic such as some soft waters and mine waters, or brackish such as salt waters and brines. The iron corrodes leaving behind a residue of graphite and iron oxides which is soft, porous and of low mechanical strength. There is little evidence of attack from the external appearance since the original dimensions and shape are retained. The fact that attack has occurred can be shown by the fact that the graphitised part is easily cut with a knife and readily marks paper. When a cross section is taken the presence of a graphitised layer is apparent (Pl. 13).

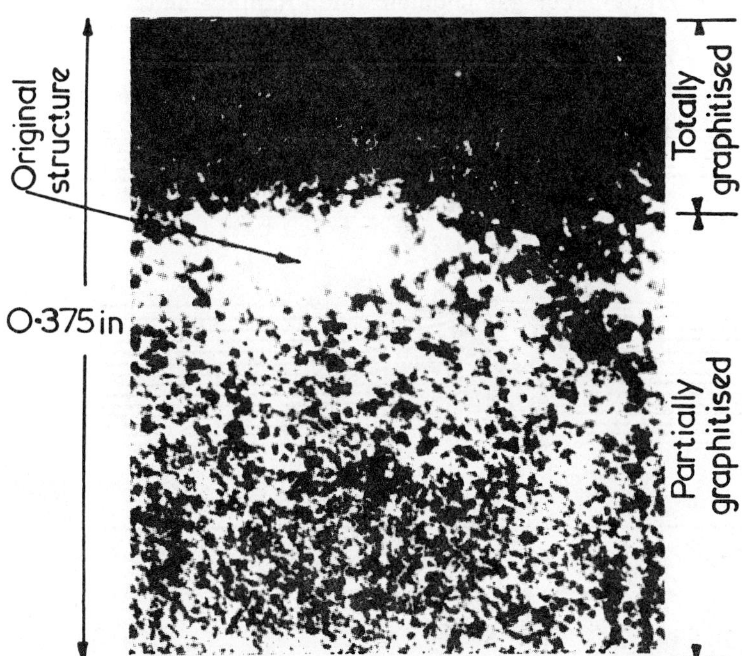

Plate 13. GRAPHITISATION OF CAST IRON WATER MAIN. 70 YEARS.

Galvanic effects can arise from contact with other metals. Although the original iron may be compatible, graphite with a very noble potential can cause accelerated attack on most metals. As we should expect, white cast iron, having no free graphite, is immune to this form of attack.

Graphitisation can be prevented to some extent by the addition of small quantities of nickel, chromium, copper or molybdenum which reduce the size and amount of the graphite flakes. Cast irons with a high nickel content undergo less attack from this cause owing to the decrease in galvanic action between the graphite and the matrix. With certain exceptions austenitic cast iron and high-silicon irons are immune to graphitisation.

CORROSION-EROSION

In waters flowing at a sufficiently high velocity for appreciable turbulence to be created, intense localised corrosion may occur by the combined action of corrosion and mechanical abrasion. This is known as <u>corrosion-erosion</u> (see also ChapVII) and includes both <u>impingement</u> attack and <u>cavitation</u>.

IMPINGEMENT

Impingement attack is caused by local breakdown of a protective layer by suspended solid particles, or gas bubbles which impinge on the surface, or

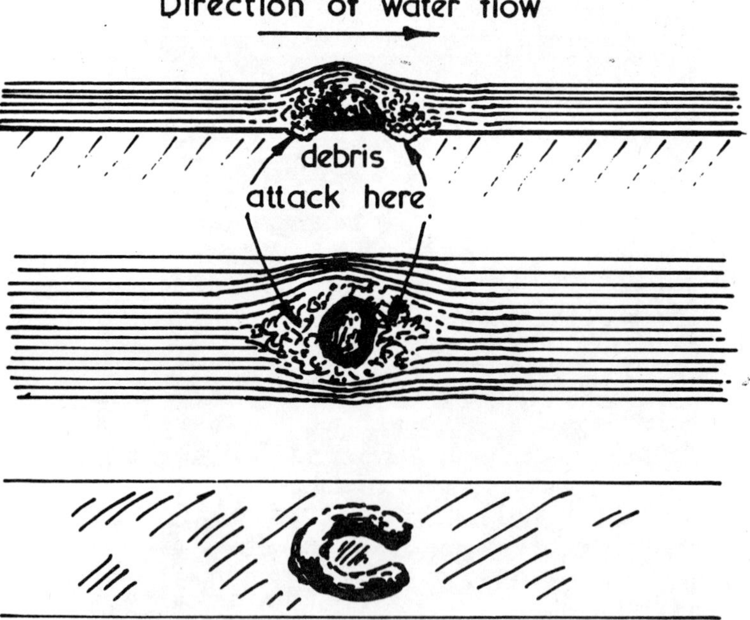

Fig. 16.   DIAGRAMMATIC REPRESENTATION OF IMPINGE-
MENT ATTACK

by turbulence alone. Such conditions arise at the entrance to pipes, at
sharp bends, near deposits and where the cross-section of the flow
stream changes abruptly. Large gas bubbles striking the surface with
force break up into smaller bubbles and release sufficient energy to disrupt any film or layer of corrosion product. Such impacts often keep
occurring at the same point on the surface and give rise to pits. These
are characterised by their sharp edged appearance, freedom from corrosion product and usually show undercutting of the metal on the downstream
side. The horse-shoe shape frequently shown is attributed to turbulence
produced by a particle on the surface. They are oriented to give the im-
pression of a 'horse walking upstream' (Fig. 16).

Metals susceptible to corrosion-erosion are usually those which form
corrosion products with very poor adhesion. The alloys more resistant to
impingement, such as aluminium-brass and cupro-nickel rely on the formation of a strong self-healing film. In condenser attack the presence of
iron appears to be an advantage whether it is added as an alloying element
to the cupro-nickels or deposited as the corrosion product from sacrificial anodes used in water boxes. The iron is an essential factor in the
formation of a stable film.

CAVITATION

Cavitation is a very serious form of severe and rapid wastage which
occurs under conditions of turbulent flow. It is caused by the formation
and collapse of vapour-filled cavities by sudden changes in pressure and
by vibration. Accordingly it is found on ships' propellers, in pumps, in
valves and on the liners of diesel engines. It is characterised by a network of sharply-outlined pits, often hemispherical, and almost completely
free from any corrosion product.

There is no sharp dividing line between impingement and cavitation attack
although we may say that the former is produced by the break-up of
large entrained gas bubbles while the latter is caused by the implosion of
vapour-filled cavities. Between these two a wide range of mixed phenomena can occur.

The main cause of the damage is mechanical but the impact of the liquid
is able to keep the affected surface free from any product or protective
film and hence plays a depolarising role. The fact that cathodic protection
and inhibitors have a beneficial effect may be interpreted either in terms
of the electrochemical effect on the corrosion or in terms of the improvement in the mechanical properties of the modified surface (Chap. XI).

CRACKING

The development of cracks leading to the failure of metal components
regularly occurs with certain alloys. In this, the amount of metal removed by corrosion is very small indeed and may be hardly measurable.
These cracks may be <u>intergranular</u>, <u>transgranular</u> or a mixture of the two
(Chap. VII).

Intergranular corrosion is associated with a difference in the corrosion
potential of the grain boundary or the grain boundary region and the rest
of the grain. The grain boundary region is anodic and attack can be very
severe owing to the large cathode to anode area. This occurs, for example,
in the case of certain stainless steels which are metallurgically unstable
in the temperature range 400-900°C. The attack in even mildly corrosive
media is so severe that the steel disintegrates into separate grains. It is
not always obvious on visual examination but the steel will usually have
lost its metallic ring when struck with a metal object and will crack on
bending, the so-called "ringing and bend test". This form of attack can
occur following welding of annealed 18/8 stainless steel. On exposure to
a corrosive medium intergranular attack will take place on either side of
the weld, the weld itself and the rest of the metal being relatively unaffected. This form of attack will be discussed in Chap. VII in greater detail.

Cracking usually occurs owing to the conjoint action of corrosion and
mechanical stress. When tensile static stresses are applied to a metal, or
residual stresses exist within the metal, the resulting cracking is referred
to as stress-corrosion cracking (Pl. 14 and 15). The cracks develop at
right angles to the applied stress which must be tensile. Aluminium, copper and manganese alloys and stainless alloys are susceptible to this form
of attack.

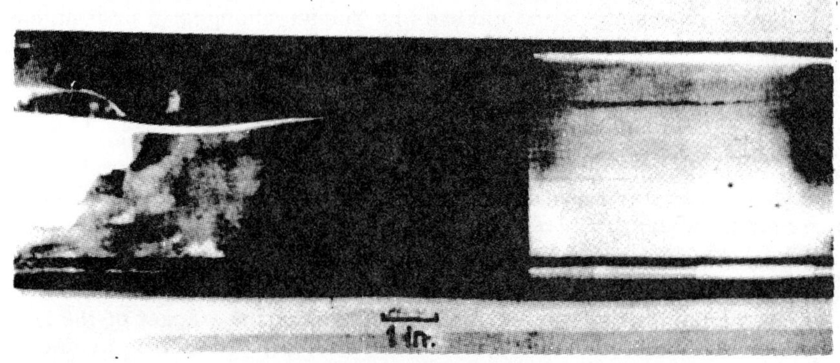

Plate 14. STRESS CORROSION CRACKING OF BRASS
SHUTTER. CHLORINATED SOFT WATER.

In the case of brass, in particular, the cracking resulting from the action
of corrosion and internal stress is usually referred to as season-cracking
from the resemblance of the cracks to those in seasoned wood. The
cracks may follow a zig-zag pattern in a piece of cold-worked metal, such
as a drawn cup, while hard-drawn brass tubing may split open longitudi-
nally.

When the stress applied to the metal is a cyclic one, the cracking resulting from the combined action with corrosion is referred to as <u>corrosion
fatigue</u>. Fracture of the metal takes place at a stress far below the
fatigue limit in air. This form of attack is not difficult to recognise since
after cleaning, fatigue cracks may usually be seen at the bottom of corrosion pits. These pits may be formed prior to, or at the same time as, the
application of stress; in the latter case the pitting rate is more rapid. In a
subsequent stage the behaviour is the same as in the case of fatigue of a
notched specimen. The crack at the base of the pit develops rapidly and the
section is gradually penetrated until fracture takes place. As in ordinary
fatigue the cracks are predominantly transgranular.

Careful examination of a cracked specimen can be very informative. Let
us consider a metal that has failed by corrosion fatigue. This usually has
two zones one of which is smooth and the other fibrous or crystalline
(Fig. 17). The smooth zone, where the crack progresses very slowly
across the section, is due to corrosion fatigue proper. When the cross-
sectional area is small enough an ordinary tensile break occurs and is
represented by the fibrous or crystalline zone. If the stressing has been
intermittent the smooth zone is likely to show ridges where penetration
stopped during the rest periods.

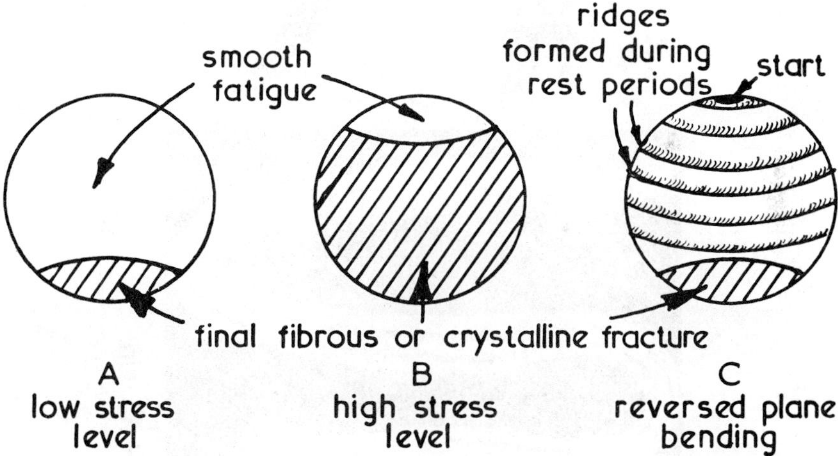

Fig. 17. CORROSION FATIGUE AT HIGH AND LOW STRESS
LEVELS

If the stress is low, corrosion fatigue cracks will penetrate the metal
until the remaining area is small, i.e. there will be a large smooth area
and a small fibrous or crystalline area. Conversely if the stress is high
the relative sizes of the two areas will be reversed.

With these types of failure, examination under the microscope is useful
in order to identify the nature of the crack. The transcrystalline crack

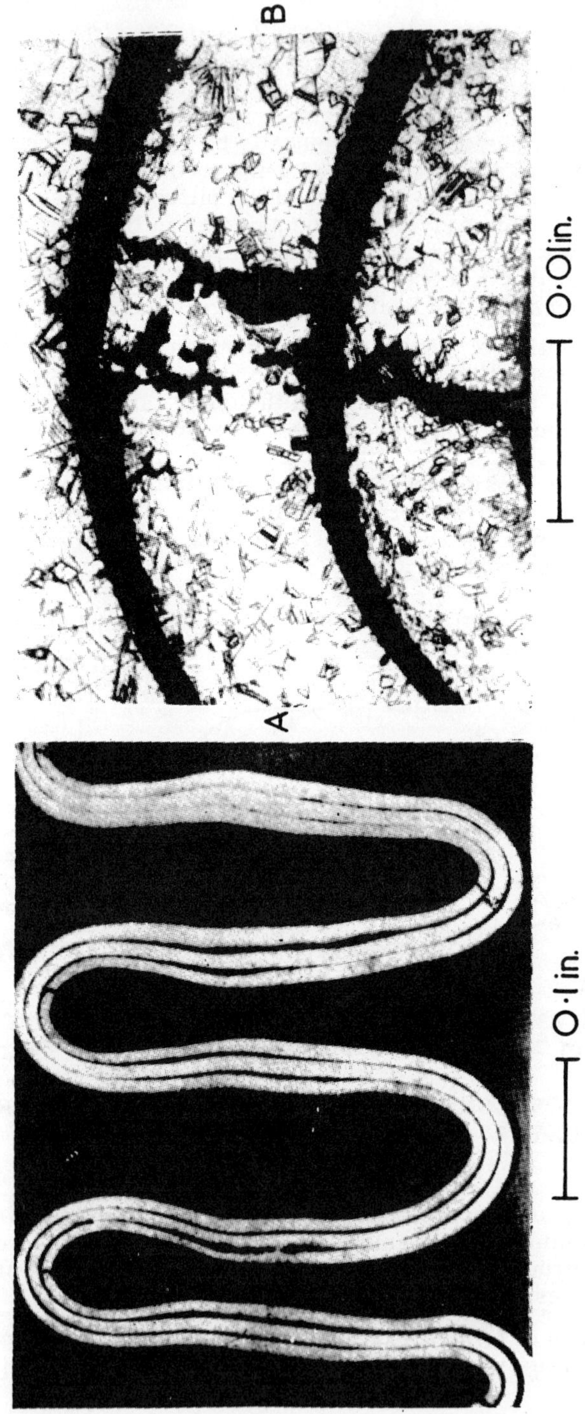

B
A
0·01 in.
0·1 in.

Plate 15.   STRESS CORROSION OF BRASS BELLOWS IN STEAM CONDENSATE.

Fig. 18.   INTERCRYSTALLINE AND TRANSCRYSTALLINE CRACKING

is straighter and blunt-ended while the intercrystalline crack wanders about between the grains and often has numerous branches (Fig. 18).

Often the type of cracking is referred to according to the agent responsible. Thus caustic cracking, also referred to as caustic embrittlement, is a type of intercrystalline attack found in boilers where it occurs in joints, seams or crevices into which water can leak and become sufficiently concentrated for attack to be initiated by the caustic alkali. High tensile stresses are essential and a common region of attack is along a line of rivets.

Hydrogen cracking, embrittlement or blistering occurs at both low and high temperatures. This active surface condition can be created by pickling. Certain poisons, e.g. arsenic, hydrogen sulphide or selenium can prevent the evolution of hydrogen. If this happens atomic hydrogen accumulates in the surface of metal, particularly in holes and inclusions, and causes blistering of soft steel. Harder steels contain many micro-cracks and the atomic hydrogen fills these and causes the cracks to spread. This trouble is of concern to the gas and oil industries. In the gas industry the presence of H_2S, NH_3, and HCN in condensate results in an accumulation of hydrogen. In sour oil the water in the oil contains acids which are chiefly H_2S and CO_2. This can result in the production of blisters on the outside of the oil pipe, i.e. on the air side. The blisters are caused by hydrogen diffusing through the pipe wall.

At high temperatures, hydrogen embrittlement is fortunately not very common, but it is caused by hydrogen produced by the reaction of caustic soda and iron. The hydrogen penetrates the grain boundaries giving rise to disintegration of the metal.

Nitrate cracking is a form of intergranular attack that takes place on mild steel at higher temperatures in the presence of nitrate and acidity. Not all mild steels are susceptible but many resistant steels may be made so by long heat-treatment. Quite commonly the steels which crack have well-defined grain boundaries.

Chapter IV

IRON AND ITS ALLOYS

Iron and alloys based on iron are by far the most widely used constructional materials. This is due not only to their excellent mechanical properties and ease of fabrication but also to their relative cheapness because
of the wide availability of suitable ores and the ease of extraction. If we
take the cost of mild steel as unity the relative costs of some other structural metals are: lead or zinc 3, aluminium 5, copper 11, nickel 15 or tin
35 (June, 1965). However, in most environments, the corrosion resistance
of iron is low compared with most other metals. This is due to a number
of factors including the ease with which cathodic reactions can proceed
on its surface. the readiness with which concentration cells are formed
and the poor protection afforded by corrosion products. The last feature
is linked with the relatively high solubility of ferrous hydroxide, a primary product of corrosive attack. Even metals such as zinc and aluminium which have a much more negative potential have a much better
resistance to attack. This is due to the restraining influence of the corrosion product on zinc and the marked tendency of aluminium to passivate.

Iron and its alloys can exhibit two entirely different types of corrosion
behaviour according to whether they are in the active or passive state. In
the active or corroding state the rate of attack varies between wide limits
according to the environment and may be very high. In the passive state
the corrosion rate is negligibly small. In natural waters under practical
conditions, iron or steel are rarely in the passive state since a large
oxygen concentration or a large supply of oxygen to the metal surface is
required in order to keep the thin protective oxide film in a permanent
state of repair. In the case of the stainless alloys, i.e. alloys of iron containing not less than 12 per cent chromium, on the other hand, passivation
readily occurs in aerated waters or waters containing oxygen. Where the
continuous repair of this film is not possible, e.g. in a crevice or beneath
grease or dirt, the corrosion behaviour of these materials does not differ
from that of iron or steel. Therefore, any discussion of the corrosion
behaviour of iron and steel will apply equally well to the so-called
"stainless" alloys in the active state.

IRON AND STEEL

On exposure to dry air an iron surface will develop a thin invisible oxide
film up to 50 Å thick. When the metal is subsequently immersed in water
this film usually breaks down locally and a number of corrosion cells
develop. The anodic and cathodic reactions have been fully described in

Chapter I. In aerated natural waters oxygen reduction is the predominant primary reaction of the cathode. As described, the products formed are hydrated ferric oxides and magnetite. None of these products of attack provide any marked restricting influence on the electrochemical process; their adhesion to the surface and their protective action are much weaker than that provided by calcium carbonate. Thus the corrosion of iron in distilled water and soft waters does not decrease appreciably with time. In this respect the protective action of rust on iron is considerably weaker than the corrosion products on lead, aluminium, tin or zinc.

In solutions with a pH value of about four where the concentrations of dissolved molecular oxygen and hydrogen ions are comparable, perceptible hydrogen evolution occurs and with decrease in pH below this figure, i.e. increase in hydrogen ion concentration, the hydrogen evolution type of attack becomes increasingly important compared with the oxygen type of attack. pH values as low as this are, however, only encountered in mine waters and in water saturated in carbon dioxide. Nevertheless even in oxygenated near-neutral and slightly alkaline solutions a certain percentage of the overall corrosion reaction is attributable to hydrogen evolution at the cathode.

This evolution of hydrogen occasionally gives rise to some concern in the corrosion of iron and steel in closed-circuit heating systems. On venting radiators the presence of an inflammable gas may be demonstrated. This is due to the ability of iron to liberate hydrogen from water. The rate at which this occurs is hardly detectable at ordinary temperatures but may readily be detected at the higher temperatures existing in hot waters. In closed systems the small amount of oxygen present initially in the water is soon consumed and subsequently a slow cathodic evolution of hydrogen takes place. Simultaneously a film of magnetite Fe_3O_4, is formed on the metal surface which eventually stifles further reaction. We know of no case in which this slow form of attack has persisted for a period longer than two years.

Even in the absence of air, corrosion of iron can take place in waters if sulphate-reducing bacteria are active. These bacteria are present in most waters and providing the sulphate concentration is adequate they will proliferate in regions where the oxygen concentration is sufficiently low, e.g. beneath corrosion deposits or in the cooler parts of closed hot-water systems (Pl. 8). The bacteria are able to use the cathodic hydrogen which would otherwise slow down or stop the corrosion process.

Cathodic reactions involving reduction of ferric ions are significant in solutions containing appreciable amounts of ferric sulphate, e.g. mine waters. The ferric ions are regenerated by oxidation of the ferrous ions formed in reaction 4, Fig. 2 by oxygen present in the solution.

INFLUENCE OF DISSOLVED GASES

Oxygen and carbon dioxide are the most important gases dissolved by waters from the air. Oxygen is an effective cathodic depolariser and increases the corrosion rate of many metals such as iron. At high concentrations, however, oxygen has a passivating action, essentially an anodic process. This dual action of oxygen, as a corrosion stimulator at the cathode and inhibitor at the anode, means that the corrosion rate at different levels of oxygen concentration particularly in the absence of aggressive ions such as chloride, shows a maximum (Chap. VIII). A similar effect is obtained by increasing the oxygen supply to the surface by increasing the amount of movement. This is reflected in the behaviour of the corrosion potential (Table 5).

TABLE 5

VARIATION OF THE POTENTIAL OF IRON WITH OXYGEN SUPPLY.

Conditions	Potential, volts. (v. NHE)
No air or oxygen	—0.53
Air or oxygen present,	
(a) stagnant solution	—0.40
(b) small rate of movement	—0.36
(c) rapid rate of movement	—0.03

In the most passive state the potential may be as high as $+1.0$ volts.

In solutions containing appreciable amounts of chloride there is no maximum and the corrosion rate increases with increasing oxygen concentration.

In cases where the oxygen concentration differs from one part of the surface to another, the part where the oxygen concentration is the higher becomes the cathode, i.e. reduction of oxygen takes place here more readily, while accelerated attack takes place on areas deficient in oxygen. Such 'differential aeration cells' can be formed by deposits or particularly by crevices. They are especially troublesome in crevices involving the stainless alloys where the passive metal in the crevice becomes active owing to the inadequate supply of oxygen. The resulting rate of attack can be very high owing to the very small area of the anode and the large area of the cathode metal outside the crevice. In very conductive solutions containing appreciable amounts of aggressive ions, e.g. sea water, unequal aeration is not a basic factor in the development of corrosion.

Carbon dioxide has no specific influence on the corrosion of iron since it does not participate in the cathodic reaction. It can, however, influence the corrosion rate owing to its acidic character and its ability to increase the hydrogen evolution type of attack by decreasing the pH of the solution. A second indirect influence is due to the fact that with increase in carbon dioxide concentration protective calcium carbonate is not formed, but remains in solution in the form of calcium bicarbonate (Chap. II). Troubles due to acidity produced by carbon dioxide are frequently encountered in condensate lines particularly where the boiler water has been treated with carbonate or where the feed water is high in carbon dioxide or sodium bicarbonate.

INFLUENCE OF pH

The influence of pH on the corrosion behaviour of iron is very character-istic. In acid solution iron dissolves freely and evenly with evolution of hydrogen. With increase in pH, i.e. decrease in acidity and hydrogen ion concentration, the corrosion rate decreases regularly until a pH value of about 4.0 is reached when the influence of oxygen reduction at the cathode begins to be appreciable. In the pH range 4.0-9.5 the rate of attack is sensibly constant. This is generally attributed to the buffering action on the solution of the product formed to give a pH at the metal surface in the range 8.5-9.5. With further increase in pH value the corrosion rate decreases more sharply to a minimum at about pH 12.0. On further increase in the pH value the corrosion rate starts to increase again with the formation of ferroates. In this increase in rate above pH 12.0 iron has a similarity to amphoteric metals such as aluminium, zinc and lead (see p. 15.).

In most natural waters the corrosion rate will be very similar as far as the pH is concerned. Although the overall attack may be reduced by increasing the pH value there is a real danger that the attack will be concentrated at a limited number of areas and the depth of pitting may become very severe, particularly in the presence of aggressive ions such as chlorides.

INFLUENCE OF DISSOLVED SALTS

The influence of dissolved salts is complex: it will depend not only on the nature of the salt but also on other conditions obtaining at the time.

In distilled water in stagnant conditions under air, the overall corrosion rate is controlled by the oxygen supply to the metal surface, i.e. diffusion and/or convection through the water and, as a result, addition of chloride, sulphate or other aggressive ion has no effect. However, if the oxygen supply is increased, e.g. when the solution is under oxygen pressure or there is movement, increase of chloride or sulphate increases the cor-rosion rate.

Chloride and sulphate ions interfere with the development of protective films and also break down any passive films more readily. Although under certain conditions the overall amount of metal lost is not markedly influenced by the concentration of chloride and sulphate the attack differs from that, say, in distilled water. In particular, the attack is more loca-lised and, as a result, pits are deeper.

In natural waters the corrosion is further complicated by the presence of other ions, e.g. calcium and bicarbonate, which have inhibitive properties. These will tend to limit the attack and cause the rate to decrease to a very low figure by the formation of an adherent carbonate-containing rust which polarises the anodic areas.

The pattern of corrosion is governed by the combination of all these factors and may vary with time.

The influence of the relative balance of carbonate and chloride is imperfectly understood. Qualitatively it is known that chloride can interfere
with the deposition of protective layers of carbonate if present in sufficient
amount. This leads to the formation of rust nodules beneath which intense
attack and pitting can take place. Various rough guides and rule-of-thumb
methods must be used in assessing whether a given water will have particularly aggressive tendencies. If the chloride and sulphate contents are
low the Langelier index is a good guide (Chap. II). At higher concentrations
of aggressive ions a hardness/chloride ratio of at least 2 is desirable.

Organic matter also has an important influence. Organic acids arising
from the decomposition and leaching of dead vegetation produces waters
with a lower pH value and hence an increased danger from corrosion.

Living organisms are responsible for fouling in sea water while in fresh
waters algae and slimes can give rise to troubles. Sulphate-reducing
bacteria have already been mentioned.

Organic matter in small concentration may have a pronounced influence
on the nature of the deposited carbonate film (Chap. II). This difference
in the product is encountered both with fresh waters and with sea water
and is, to say the least, imperfectly understood. The materials responsible are difficult to isolate owing to their very low concentrations, much
less than one ppm.

It will now be apparent that the use of corrosion rates for steel in natural
waters can be dangerous if the conditions are not fully appreciated. Very
little steel is used without some protection but some assessment of corrosion has to be made for the period between breakdown of the protection
and repair. The rate usually falls with time due to the restraining action
of the product and deposited scale and in quiescent soft fresh water the
rate may be 15 to 20 mdd while in hard water the figure may be as low as
2.5 mdd. In sea water and estuarine waters pitting is the danger and
cannot be predicted. The following figures are actual records of pit depths
and are, therefore, given in mpy as well as mdd. An average rate of corrosion of mild steel immersed in quiet sea water is 5 mpy (27 mdd). With
half-tide immersion this figure will be multiplied by a factor of between
2 and 5. Maximum rates of pitting in sea water are much higher than the
average figure given and may be as high as 17 mpy (93 mdd). This may
also be the sort of figure obtained in a tropical estuarine water. The corrosion rates increase with movement, e.g. at 6.6 ft/sec the average rate
is about 25 mpy (137 mdd) and at 33 ft/sec about 70 mpy (384 mdd). The
maximum pitting rate at 33 ft/sec might be as high as 160 mpy (877 mdd)
which would mean that a $1/4$ in. plate would be perforated in 1.6 years if
the rate were maintained as against 3.7 years at the average rate.

INFLUENCE OF MILL-SCALE

Although the presence of mill-scale on the metal surface does not affect
the overall corrosion it markedly increases the localisation of attack

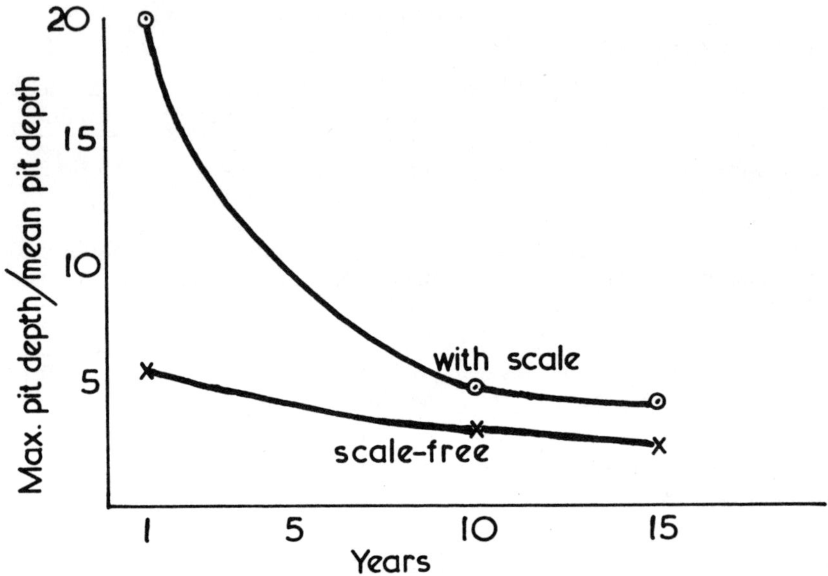

Fig. 19.   EFFECT OF MILL-SCALE ON PITTING OF STEEL
IN SEA WATER

(Pl. 3 to 6). Its influence is clearly demonstrated by the results of sea
water immersion tests over periods up to 15 years (Fig. 19). With steel
free from mill-scale the pitting factor decreases slowly with time indicating a gradual spread of attack. With steel originally covered with mill-
scale the penetration after one year was four times as great and although
the difference decreases, the effect is still evident after 15 years.

IRONS

CAST IRON

By virtue of its comparative cheapness and ease of manufacture, cast
iron has always been used extensively for pipes and castings. Cast irons
are ferrous alloys containing more than 1.7 per cent carbon and in the
normal slowly-cooled cast form exhibit the familiar grey fracture due to
the presence of free graphite. This graphite, in the form of flakes, is the
cause of the brittleness of cast iron and is the important difference from
mild steel. Grey cast iron consists of graphite, pearlite, and either ferrite
or cementite, Fe_3C, depending on whether the combined carbon is below or
above the eutectoid composition. The ferrite is basically α-iron but may
contain nickel, manganese, silicon, phosphorus and possibly some carbon
in solid solution. The solubility of carbon in ferrite is only 0.025 per cent
at the eutectoid temperature and is considerably less at lower tempera-

tures. Rapid cooling of a casting prevents the graphite from separating and gives rise to the very hard white iron with free Fe_3C and pearlite. On annealing white iron, the carbon separates in a nodular form and the metal becomes ductile. The common malleable irons are Blackheart in which annealing is carried out in an inert material and all the carbide is decomposed to give a structure of nodular carbon in ferrite, and Whiteheart, in which annealing is carried out in decarburising conditions in fine iron ore and the structure retains varying amounts of pearlite.

Small percentages of alloying additions have little effect on the corrosion resistance of cast iron but larger additions have been made to develop cast irons of better mechanical and corrosion resistant properties. These include the high silicon irons with up to 18 per cent silicon, the high chromium irons with 25 to 35 per cent chromium and the austenitic irons of the Ni-resist type containing not less than 15 per cent nickel and 7 per cent copper. These alloys are discussed later in this chapter.

The initial corroding characteristics of cast iron are modified to some extent by the presence of a casting skin which may be either white iron, if there has been any chilling, or high in silica if moulded in sand. Cast iron may contain the following phases arranged in descending order of potential from the most noble to the least noble, viz: graphite, Fe_3C, Fe_3P, MnS, FeS and ferrite. The difference of potential depends on the corroding medium but between graphite and ferrite it is of the order of 1.0 volt with Fe_3C and Fe_3P approximately mid-way between.

The effect of the constituents of grey cast iron on the initial corrosion in a hard mains water with and without the addition of 1.5 per cent sodium benzoate is shown in Pl.16. In water alone, the initial sites are rather random, but the presence of benzoate is striking in that the attack starts around the phosphide eutectic which itself is etched.

If heat treatment leads to an increase in the heterogeneity of the structure the number of possible micro-cells will increase. Although these have an effect in acid solution their influence on the corrosion resistance in neutral and near-neutral media is negligible. The corrosion resistance of the plain and low-alloy cast irons in natural waters is neither influenced by the type of iron nor even by whether the graphite is in the flake or spheroidal form. All the cast irons behave quite well in distilled or fresh waters, but the rate of attack increases with salt concentration, aeration and temperature. In distilled water at 16°C the corrosion rate of grey cast iron is 55 mdd.

In soft acidic waters and sea water graphitisation of the cast iron occurs leaving a porous graphite residue impregnated with insoluble corrosion products. The corrosion rates for cast iron and steel are not very different, the apparent better behaviour of cast iron under some conditions being due to its normally much thicker section and to the graphitisation which retains the original form and enables pipes to retain water. The main disadvantage of this form of attack is that the affected pipe loses its original strength and is easily fractured by earth movement. A pipe which was removed after 78 years was still carrying land drainage water but was

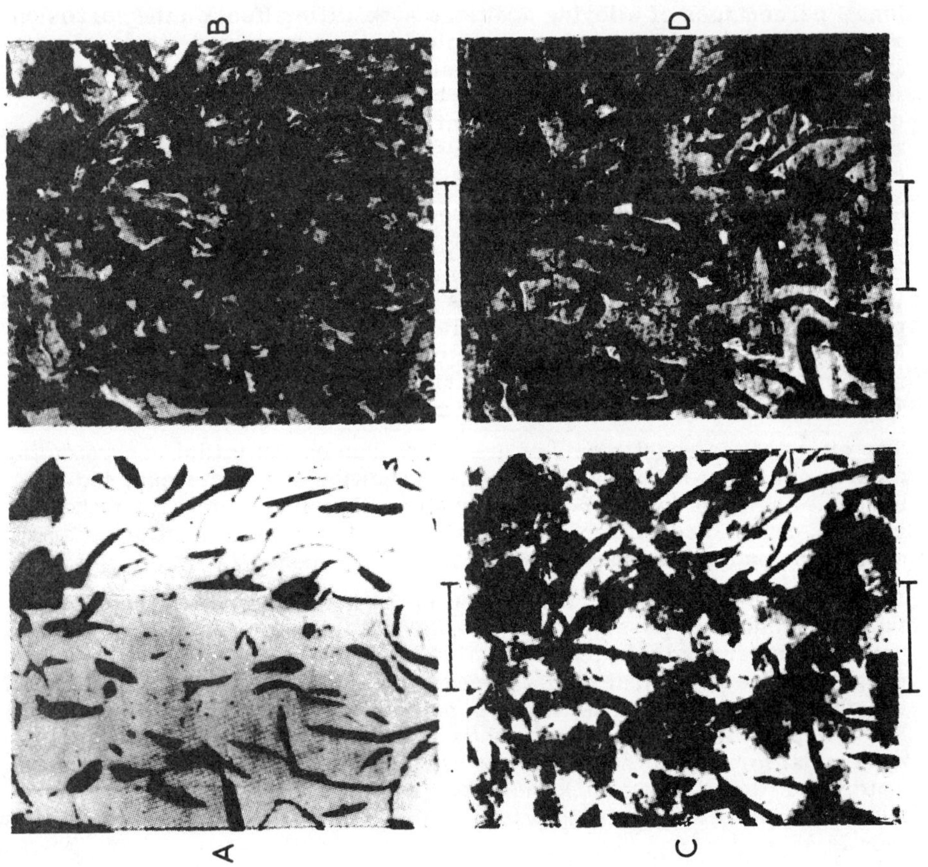

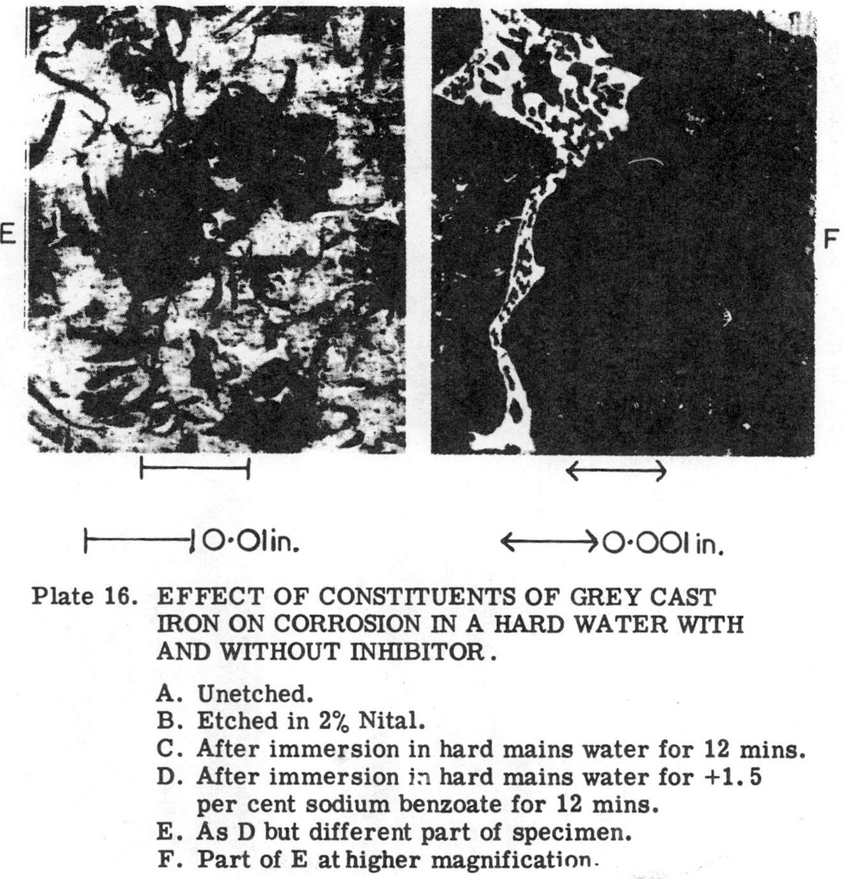

Plate 16. EFFECT OF CONSTITUENTS OF GREY CAST IRON ON CORROSION IN A HARD WATER WITH AND WITHOUT INHIBITOR.

A. Unetched.
B. Etched in 2% Nital.
C. After immersion in hard mains water for 12 mins.
D. After immersion in hard mains water for +1.5 per cent sodium benzoate for 12 mins.
E. As D but different part of specimen.
F. Part of E at higher magnification.

completely graphitised (Pl. 13). This loss of strength can be avoided to a
certain extent by reducing the size of the graphite flakes or by alloying
with small amounts of nickel, chromium, copper or molybdenum. Graphitisation can have serious consequences owing to the formation of galvanic cells between the graphitised iron and any dissimilar metal in contact
with it.

In stagnant sea water the corrosion rate of grey cast iron is of the order
of 25 to 30 mdd but this rate is increased by movement and long-term
immersion figures are around 55 mdd. Speeds up to 30 ft/sec will increase
the rate still further to about 250 to 300 mdd.

WROUGHT IRON

As little or no wrought iron is produced now its behaviour is largely of
historic interest. Nevertheless old wrought-iron pipes are still in use and
when the need arises are replaced by mild steel. The comparison of
service life often favours the wrought iron and it is well to appreciate the
reasons for the difference.

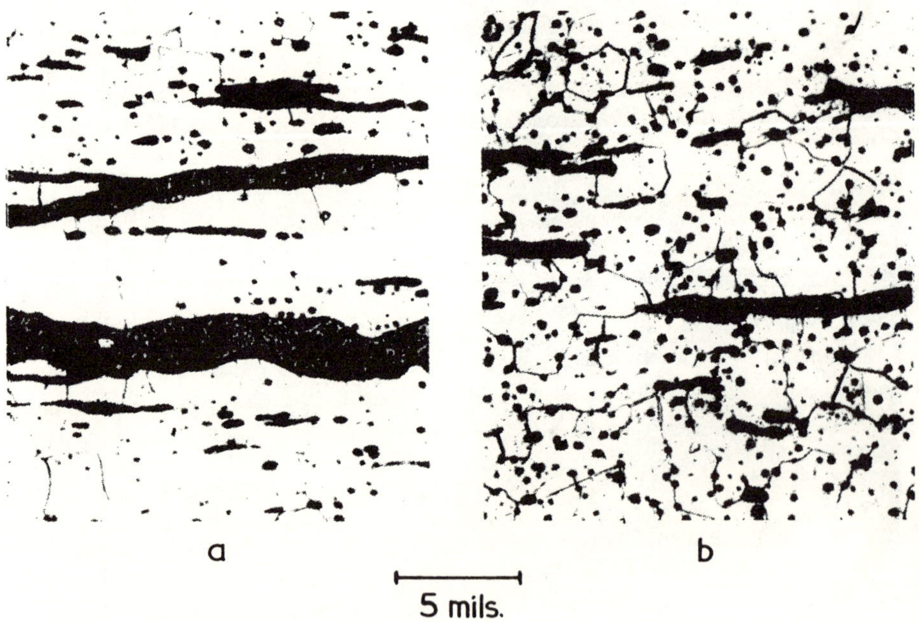

Plate 17. MICROSECTIONS OF WROUGHT IRON.

(a) longitudinal and (b) transverse.

Wrought iron consists essentially of a fairly pure iron with slag inclusions (Pl. 17). the percentage of slag by weight varying from 1.0 to 4.0

per cent but usually being about 2.0 per cent. The notable difference between wrought iron and mild steel is, therefore, the presence of slag in the former. Wrought iron has less carbon and manganese and usually more phosphorus. The silicon content is of the same order in the two materials, i.e. between 0.10 and 0.20 per cent, but with this important difference—in steel the silicon is almost entirely in solid solution and dispersed homogeneously, but in wrought iron there is virtually none in the basis metal, the whole being in the slag as a complex iron silicate.

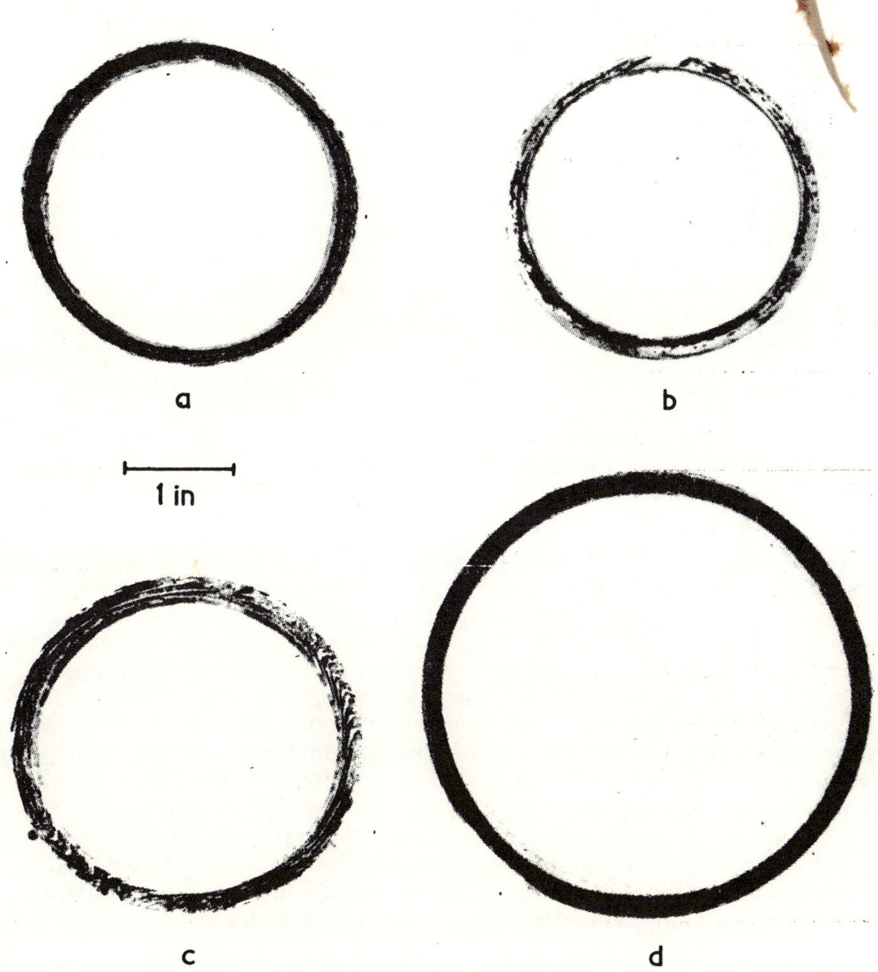

Plate 18. DEEPLY ETCHED SECTIONS OF WROUGHT IRON (a, b, c,) AND MILD STEEL (d) TUBES.

a and b, lap-welded; c, hot-extruded.

The heterogeneity of wrought iron is shown by the different rates of attack on different zones when etched in strong acid. Sections of lap-welded wrought-iron boiler tubes show a number of resistant zones parallel to the surface except where they form an outcrop at the weld (Pl. 18 a and b). Hot-extruded wrought-iron tubes also show resistant zones but these are irregularly disposed in relation to the tube surface and the outcrops are numerous (Pl. 18c). Mild steel on the other hand is etched evenly over the section (Pl. 18d).

Two types of resistant zones have been shown to exist in wrought iron, the more resistant containing appreciable amounts of copper and nickel as alloying elements. The pitting of wrought iron is considered to proceed as follows. In a direction parallel to the rolling direction or at 90° to it, but in the plane of the sheet, attack can proceed in the non-resistant zones to an extent similar to that for pure iron or steel. Perpendicular to the plane of the sheet, however, pitting will proceed until a resistant zone is reached when the rate of penetration will be considerably reduced and the attack will be diverted sideways. Corrosion will then only proceed at an appreciable rate in a direction perpendicular to the surface where there are imperfections in the resistant zones. Thus, provided that these zones are encountered frequently enough, perforation of the sheet may be delayed or prevented.

Although an outcrop occurs at the weld, the overlap at the weld is sufficient to ensure that the susceptibility to perforation in this region is slight. This would not apply to the hot-extruded wrought-iron tubes where there were numerous outcrops (Pl. 18c). These tubes are also mechanically poor and split on attempting to expand them into the end-plates of a boiler.

The reported longer life of wrought-iron tubes compared to mild steel ones is not entirely attributable to these resistant zones but is also linked to a difference in the nature of the scale and its adhesion to the metal surface. The sideways deflection of the pits in wrought iron leads to undercutting of the metal surface which will provide a good key for any scale formed. With the progress of corrosion, slag inclusions with a high silicon content will be uncovered and these can modify the type of scale formed on the tubes. A thin, very tough, adherent scale on wrought-iron tubes which had been in service for 26 years was found to contain a complex silicate of the zeolite type, probably containing calcium and magnesium. In another case the silica content was found to vary from 10 per cent in the bulk of the scale to 20 per cent or more at the metal surface. These observations strongly suggest that the silicon in the slag combines with corrosion and scale deposits to form a complex silicate.

SILICON IRON AND CHROMIUM IRON

Silicon irons, which have a maximum corrosion resistance at 14 to 15 per cent silicon, are extremely hard and brittle and are used only as castings. They were developed primarily for use with acids and their excellent resistance is attributed to the development of a silica film on the surface.

The chromium-iron alloys have a similar resistance to the silicon irons.
If justified by conditions, both alloys may be used in sea water or acid
mine waters, particularly if there is entrained sand or coal dust because
their resistance to erosion is enhanced by their hardness.

AUSTENITIC NICKEL IRONS

These alloys contain from 14 to 30 per cent nickel with various other
additions such as copper, chromium and molybdenum and were developed
for the chemical industry for resistance to acids. They are also much
more resistant to natural waters than unalloyed iron and may be used
with advantage if the greater initial cost can be justified. In distilled
water the corrosion rate of austenitic iron is 3.3 mdd and in sea water
11.0 mdd as compared with a rate of 55.0 mdd for grey cast iron in
either water. The austenitic irons have a more noble potential than cast
iron and in highly conducting solutions they are attacked less than unalloyed iron in contact with bronze because of the smaller current flow.

NICKEL-IRON ALLOYS

Alloys of iron and nickel, typical materials containing 26 or 36 per cent
nickel, are used mainly on account of their physical properties rather
than their corrosion resistance. Their corrosion resistance is better
than iron or mild steel in mains water, salt water and sea water but the
attack is still appreciable and they are less resistant than pure nickel.
The rate of attack fully immersed in sea water is of the order of 10 mdd
compared with 25 mdd on mild steel. Unlike mild steel, the corrosion
rate is lower under alternate immersion conditions, e.g. half-tide, than
under continuous immersion. In both conditions, while there may be a considerable degree of general attack there is no tendency to deep pitting.

In brackish waters a relatively high corrosion rate may occur owing to
contamination by organic acids derived from vegetation.

STEELS

LOW ALLOY STEELS

Additions of small amounts of a wide variety of other elements are often
made to low carbon and medium carbon steels with the object of improving their mechanical properties. The percentage of such elements may
range up to 2 to 3 per cent and occasionally as high as 5 per cent.

Of the elements always present in steels, sulphur and phosphorus are
undesirable and the usual specifications tolerate no more than 0.04 per
cent of either. High sulphur gives rise to 'hot-short' steels, i.e. brittle at
high temperature, while high phosphorus gives rise to 'cold-short' steels,
i.e. brittle at low temperature. Silicon is added as a deoxidiser and is
present in amounts between 0.1 and 0.2 per cent. Manganese is present
between approximately 0.4 and 0.6 per cent but larger additions confer
the property of self-hardening on alloy steels, e.g. for steel rails. Manga-

nese readily combines with sulphur to form manganese sulphide, the excess going partly into solid solution in the ferrite and partly to form manganese carbide, Mn_3C, which occurs with the cementite, Fe_3C, in the pearlite.

In the amounts normally present in mild steel these elements have little influence on its general corrosion behaviour in neutral media even though they increase the number of cathodic micro-inclusions at the grain boundaries. The corrosion rate is controlled by the transport of oxygen and the increase in the number of potential cathodes does not raise the cathodic efficiency since sufficient cathodes exist even on pure iron.

More interest is attached to the possible influence of small additions of copper, chromium and nickel. Although the effects are small and variable they are worthy of careful consideration.

The effect of copper in reducing corrosion under atmospheric conditions is not disputed but there is little evidence for any beneficial effect under immersed conditions in natural waters. Nevertheless under boiler conditions with water which becomes acid owing to the hydrolysis of magnesium salts present in the feed water, addition of small amounts of copper up to 0.3 per cent has a very marked effect (Fig. 20). Not only is the order of magnitude of the pit depth reduced but the attack is more general and even under highly oxygenated conditions perforation of the boiler tubes is prevented. This is attributed to the catalytic action of the metallic copper on the formation of a protective layer of magnetite which diverts the attack sideways. Even though such development of acidity would be prevented by the addition of alkali, danger from acid condensate or production of acid can be reduced by ensuring that the requisite amount of copper is present.

Alloying with chromium appreciably reduces the corrosion of steel in sea water. The addition of 3 per cent chromium to steel reduces the corrosion rate by about 50 per cent. If the chromium content is raised to 5 per cent, however, intense localised attack can occur.

Low alloy steels containing nickel, copper and small amounts of molybdenum in addition to chromium show enhanced stability in sea water. Copper by itself does not show any positive effect but when combined with small amounts of chromium, nickel and aluminium an increase in resistance over low-carbon steel of 25-30 per cent has been observed.

Although the question of whether and to what extent the corrosion behaviour of steel may be improved by small amounts of alloying elements is still a debatable one, there is no doubt that such additions influence the nature of the corrosion product. The rust formed on low-alloy steels is less permeable and more protective than on ordinary steels and any improvement in behaviour must be associated with this fact. Furthermore, on painted steel the more compact nature of the rust presents less danger of paint films lifting near to breaks or 'holidays'.

On the other hand the mill-scale on low-alloy steels is more firmly held and if the scale is not removed the danger of localised attack is correspondingly greater.

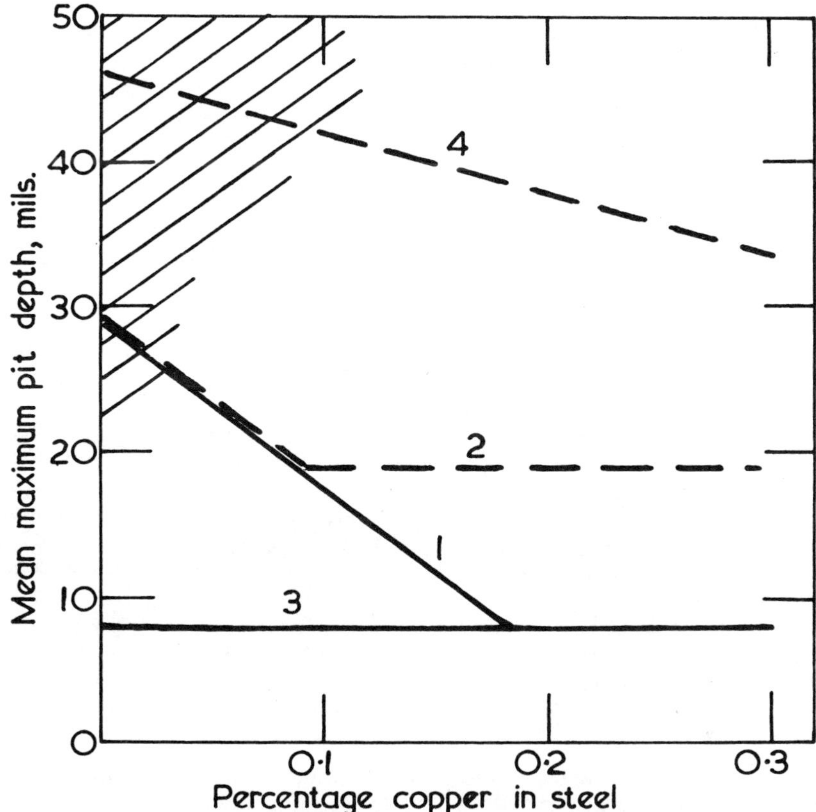

Fig. 20. INFLUENCE OF COPPER IN STEEL ON PITTING
UNDER BOILER CONDITIONS

Seamless tubes: curve 1, 4 ml/1 oxygen
                  "   2, 15 ml/1   "
                  "   3, 4-15 ml/1 oxygen, pH 8
Welded tubes: "   4, 4 ml/1 oxygen

Shaded area: Occasional tube perforations in this
region.

Although the low-alloy steels are more expensive they can be used with
advantage in certain applications. For example 3 per cent chromium
steel has been used for bolts in lock gates while a 2 per cent nickel alloy
is beneficial in the welding of mild steel. Extended trials on the use of
low-alloy plate for small vessels in polluted inland waters have shown
that the depth of pitting just above the water line is reduced by one-half
and that below the water line by one-third when compared with a low-
carbon steel.

STAINLESS STEELS

The stainless steels (Table 6) represent an important series of alloys of
iron with chromium, and chromium and nickel, which have a high resistance to many corrosive environments. In certain conditions, however,
their resistance to attack is far from good and may be no better than
mild steel. Lack of appreciation of this fact has caused many users to
take the designation 'stainless' too literally and to apply such steels in
the wrong way or to use an unsuitable type of steel.

The ready ability of the stainless steels to passivate in oxygenated environments requires the alloying addition of at least 11.5 per cent chromium.

The corrosion resistance of the stainless steels is due to the formation
of a very thin passivating film on the metal surface. Such a film forms
spontaneously on exposure of the clean metal surface to dry air or more
rapidly on immersion in an oxidising agent such as nitric acid. When such
a surface is immersed in water containing oxygen, a dynamic equilibrium
is set up between alternate breakdown and repair of the film. If, however,
such film repair is prevented by lack of oxygen access. severe corrosion
can take place at a rate comparable with that of iron or mild steel. Such conditions can arise beneath debris, barnacles or in crevices (Pl. 9). In the
use of stainless steels this behaviour should always be borne in mind and
the installation designed so as to minimise fouling of the metal surface
and where possible to provide for adequate cleaning of the surface. All
possible crevices should be eliminated by design, e.g. by replacing riveted
joints with welded ones. Where this is not possible the situation may be
improved by sealing the joints, but this method is not completely reliable.

The range of environments in which the stainless steels are able to remain passive is markedly broadened by increasing the chromium content
and to a less extent by increasing the nickel. In this connection supplementary alloying with small amounts of molybdenum is also advantageous.

When corrosion of stainless steel does occur it is usually localised. The
presence of chloride increases the susceptibility to pitting attack particularly at grain boundaries. This tendency is larger with higher carbon
contents but it can be reduced by supplementary alloying with small
amounts of molybdenum or silicon. Increase in pH decreases the number
of pits and their rate of growth.

Apart from this tendency to pitting in chloride solutions, stainless steel
is susceptible to intergranular attack and stress-corrosion cracking
(Chap. VII). The susceptibility to intergranular attack is increased by
heat treatment in the range 400-900°C and is difficult to avoid during
welding and stress relief. The sensitivity may be reduced by annealing,
by keeping the carbon content as low as possible or by the addition of
stabilising elements. Austenitic stainless steels are susceptible to stress-
corrosion cracking in steam condensate, high temperature water and,
particularly, hot sea water. In most cases chlorides are involved but
where the internal stresses are very high, as with spring-temper stainless steels, cracking can occur even when chloride is absent.

TABLE 6

STAINLESS IRON AND STEELS

Type	Designation	Percentage composition		
		C	Cr	Ni
Martensitic	(a) Stainless iron	0.07 to 0.10	12.0 to 14.0	—
	(b) High chromium stainless steel	0.22 to 0.44	12.0 to 14.0	—
	(c) High chromium low nickel stainless steel	0.10 to 0.20	16.0 to 20.0	1.0 to 3.0
Austenitic		0.05 to 0.15	15.0 to 22.0	6.0 to 11.0
Ferritic		0.05 to 0.15	16.0 to 30.0	—

Martensitic and ferric steels are magnetic but austenitic steels are not.

Austenitic and ferritic steels are not amenable to hardening.

Corrosion in natural waters. Stainless steel is normally in the passive state in a wide range of supply, lake and river waters even when pollution is relatively high and irrespective of whether the water is cold or hot. It generally remains free from rusting or pitting whether it is submerged partially immersed or subjected to alternate wet and dry conditions. Superficial rusting and pitting may occur due to incorrect heat-treatment or the formation of active concentration cells.

In strongly saline solutions and sea water the behaviour of stainless steels depends on the relative strengths of passivating influences and the tendency for the chloride to destroy the passive film. If the aeration is sufficiently strong these materials are stable but in stagnant regions where the oxygen supply is low there is likely to be marked local attack and pitting which varies with the type and composition of the stainless steel. Austenitic stainless steels containing molybdenum are the most stable whilst the straight chromium alloys are the most subject to localised attack. Steels containing molybdenum, as well as those with higher chromium and nickel contents, e.g. 25/20 chromium-nickel. are much more resistant to pitting attack than ordinary 18/8 chromium-nickel steels.

There is little advantage in using stainless steel instead of ordinary steel in stagnant sea water since fouling takes place and comparable depths of pitting are obtained. On the other hand in rapidly moving sea water fouling does not occur and, while the corrosion of mild steel is increased, the pitting of stainless steels falls markedly and the amount of attack becomes negligible. Thus, in stagnant sea water the rate of pitting on 18/9 chromium-nickel steel is about 73 mpy but when the flow velocity is in the range 240-300 ft/min the corresponding figure is only 4 mpy. Between the extremes of no movement and rapid movement the performance of the stainless steels is variable and resistant alloys containing molybdenum should be used. Alloying with titanium and niobium, in the absence of molydenum, does not increase the resistance to localised attack in sea water.

Tubular condensers of stainless steel should be designed so that the water flows through the tubes at a velocity greater than 5 ft/sec and not through the intertube region. If the velocity is low enough to give pitting conditions the tube thickness should be correspondingly greater so that the pitting has a chance to slow down or stop before perforation occurs.

Polishing the surface does not protect stainless steels from local attack in sea water. The number of pits may be less but their depth is often greater owing to the less favourable anode/cathode area.

Treated waters. The composition of boiler water is usually such that its aggressiveness to stainless steels is negligible. Occasionally, however, stress corrosion cracking of the austenitic types can occur.

Stainless steel is used in some types of atomic reactor in contact with high-purity water at high pressure and temperature in the range 260-315°C. Under such conditions the austenitic stainless steels are relatively free from attack. The influence of crevices is eliminated by making the clearance between the metal surfaces 5 mils or more.

Chapter V

CORROSION OF NON-FERROUS METALS

Although iron and steel are considered first for the majority of installations it is in only very rare cases that they can be used without some form of protection. As a class, the non-ferrous metals are more resistant to attack by natural waters owing to their self-passivation, e.g. aluminium; formation of restraining films, e.g. zinc; or their inherent nobility, e.g. copper or nickel. This means that in almost every case their use would be preferred from a corrosion point of view. However this is often impossible because of their inadequate mechanical properties but, more particularly, because of the large cost involved. Nevertheless they still play an important role in the construction of small equipment and any installation where freedom from corrosion troubles is a primary factor in design.

The strength problem can be surmounted by their use in the form of coatings, cladding or duplex tubes where the advantages of the corrosion resistance of the non-ferrous metal or alloy is combined with the strength of the basis metal.

ALUMINIUM AND ITS ALLOYS

On account of its lightness, ease of fabrication, high strength when alloyed and resistance to corrosion, aluminium has a large number of technological applications which has been increasing continuously for several decades. In contact with water and aqueous solutions it is widely used in the food and chemical industry and in the fabrication of domestic utensils and washing machines. More recent applications are as a constructional material for the hulls of boats and parts of nuclear reactors.

The corrosion behaviour of aluminium and its alloys depends on the metallurgical condition, the impurities and alloying elements present, and the nature of the dissolved salts in the water. A high resistance to attack depends on the formation of a tightly adherent and protective surface film consisting of an anhydrous or hydrated oxide. The thickness of the film formed in dry air is 150 to 200 A°, but it varies according to the conditions of formation and may be as great as 1000 A°. This <u>passivation</u> of aluminium is not only produced by reaction of the metal with oxygen in the air or dissolved in water but also by direct reaction with the water itself.

The stability of the oxide film in aerated waters is similar to that formed on chromium and only inferior to that formed on titanium. It should be noted that in aerated chloride solution under comparable conditions, pure aluminium has a lower corrosion rate than copper, a semi-noble metal.

The film formed on aluminium is generally stable in the pH range 4.5-8.5 but outside this range the aluminium becomes active, its potential becomes more negative and dissolution takes place with evolution of hydrogen. Although aluminium dissolves particularly easily in alkalis the corrosion resistance above pH 8.5 is dependent on the source of alkali and in many alkaline waters aluminium is quite stable.

A much thicker oxide film with greater corrosion and abrasion resistance may be formed by anodic treatment of aluminium in certain acid electrolytes (Chap. X). Such coatings may be from 0.1 to 1.0 mil thick and have low electrical conductivity and excellent adhesion to the basis metal. Their corrosion resistance may be further improved by boiling in water when the amorphous film is converted into the more electrically resistant γ-monohydrate, Boehmite, and the porosity is eliminated.

ALUMINIUM

The corrosion resistance of aluminium depends very much on the purity of the metal. High purity aluminium has the maximum stability and less pure material should only be used when the conditions are less demanding. The composition, location, amount, continuity and potential of the micro-constituents relative to the aluminium solid solutions are the basic factors determining the amount and distribution of corrosion.

The corrosion stability is appreciably reduced on increasing the percentage content of more electropositive metals such as copper, platinum, iron, nickel and tin. These heavy metals not only worsen the continuity of the protective oxide but also make the cathodic process easier. In neutral solutions aluminium corrodes with the oxygen-reduction type of attack but the contribution of the hydrogen-evolution type of attack increases with the amount of heavy metal. Thus with aluminium-copper alloys, e.g. duralumin, in chloride solutions, both cathodic reactions take place to a similar extent.

The action of copper is much greater than that of iron or silicon although copper has a significant solubility in aluminium and iron has a very low solubility and forms an intermetallic compound, $FeAl_3$. This is obviously explained by the occurrence of copper-aluminium corrosion cells owing to the secondary displacement of copper ions from solution. Aluminium-iron or aluminium-iron-silicon particles form micro-cathodes over which the oxide film is weak and electrochemical attack of the surrounding aluminium is promoted.

The greater corrosion resistance of high-purity aluminium is due to the decrease in number of such active points with increase in purity. With manganese, aluminium forms a compound which has almost the same potential as aluminium and hence aluminium-manganese alloys have good corrosion resistance. Similar considerations apply to alloys of aluminium with more electronegative metals such as magnesium.

In general, high-strength aluminium alloys are not as resistant to corrosion as other alloys. This has led to the development of 'alcladding' in which a core alloy, chosen for its mechanical properties, is clad with pure aluminium or aluminium alloy to which it is metallurgically bonded. The cladding alloy has not only a high inherent resistance to corrosion but is also sufficiently anodic to the core alloy to provide electrochemical protection. If attack occurs at any point it will only penetrate as far as the core alloy, or diffusion zone, where further progress is stopped by cathodic protection and corrosion then proceeds laterally along the interface.

The ability of aluminium to self-passivate makes it quite a stable material in many neutral and weakly alkaline media. However in certain solutions the breakdown in passivity leads to a very intense localised attack. Although it is stable in sulphate solutions aluminium is very sensitive to chloride ions especially in acid media. The tolerance towards chloride becomes greater with increase in purity of the aluminium but in sea water both aluminium-magnesium and aluminium-magnesium-silicon alloys have a high corrosion resistance and have been successfully used for the hulls of vessels. Contamination in harbour waters has a varied action.

Aluminium is an excellent material for dealing with high-purity waters and condensates. Here the corrosive constituents are dissolved oxygen and carbon dioxide which do not affect aluminium and the take-up of aluminium by such waters is less than 0.02 ppm. In boiling water attack takes place with hydrogen evolution and if the silicon content is very low, 0.0025 per cent or less, no decrease in rate of attack takes place. This attack can be reduced either by increasing the silicon content of the aluminium or by the addition of small amounts of silicic acid to the water.

Aluminium is sufficiently resistant for use in atomic reactors at temperatures below 150°C but rapid corrosion occurs at temperatures above 230°C. At elevated temperatures alloys containing iron and nickel, e.g. 1.0 per cent nickel and 0.5 per cent iron, have outstanding resistance. Unlike the behaviour at lower temperatures, the resistance is improved by keeping the silicon content as low as possible: when the silicon content is 0.17 per cent the rate of attack is 16 mdd but when it is less than 0.005 the attack is only 1.3 mdd.

Many natural alkaline waters are compatible but certain waters produce etching which can be reduced by the addition of sodium silicate. The entrainment of alkaline boiler compounds in steam condensate is a possible hazard and it is good practice to trap out alkaline boiler compounds upstream. The storage of steam condensate in a tank of a 1.2 per cent manganese alloy gave an aluminium content of only 0.04 ppm after 3 months.

Salts of heavy metals such as copper, lead, tin, nickel and cobalt, and to a lesser extent iron, are able to promote localised attack particularly in acid waters. Thus in mixed-metal systems in which waters circulate, increased corrosion of aluminium and its alloys is observed if components of copper or copper alloys are present even if there is no electrical contact with aluminium. To combat severe pitting in this type of fresh water clad products, e.g. the 1.2 per cent manganese alloy clad with pure aluminium, were developed.

Aluminium and its alloys are very sensitive to contact corrosion. This is most dangerous with the more electropositive metals such as copper and its alloys but troubles can also arise from contact with iron, steel and stainless steel. Zinc and cadmium are compatible with passive aluminium and even protect the latter to some extent. Contact with magnesium and magnesium alloys, although these have a more negative potential, should be avoided because strong cathodic polarisation of aluminium produces alkalinity in the water which can transform aluminium into the active state and cause discharge of hydrogen from its surface. The effect of such bimetal contacts is greater in conducting media containing chloride ions.

ALUMINIUM ALLOYS

Aluminium alloys have better mechanical properties than pure aluminium but, as a rule, a lower corrosion stability. This applies particularly to alloys with copper, to a certain extent to alloys with silicon, and to only a small extent to alloys with zinc, manganese and magnesium. The corrosion behaviour of many alloys can, however, be improved by suitable solution heat treatments.

Aluminium-Copper Alloys The first 'duralumin' containing 3.5 per cent copper and 0.5 per cent magnesium was patented by Wilm in 1910. Today, alloys of this type, containing 3.5-5.5 per cent copper and a little manganese, magnesium or silicon, are the most widely used of the high strength alloys. They may be heat treated but, owing to their ability to age-harden, local and intercrystalline corrosion are often encountered (Chap. VII). Localised corrosion is associated with the less protective character of the oxide film and in corrosive environments additional protection is required. Casting alloys containing copper are also used by owing to their low corrosion resistance their use is limited to media of low aggressivity.

Aluminium-Magnesium Alloys Alloys of aluminium and magnesium have inherently a much greater resistance to corrosion than the aluminium-copper alloys. At high temperatures aluminium can dissolve up to 15 per cent magnesium but at low temperatures the solubility is slightly less than 3 per cent. With magnesium contents up to 3 per cent the corrosion resistance is approximately the same as pure aluminium but the mechanical strength is low. Alloys containing about 5 per cent magnesium, as well as those containing magnesium and silicon, can form intermetallic compounds, Mg_2Al_3 and Mg_2Si respectively. These alloys are stronger but have a lower corrosion resistance.

Alloys containing 8-12 per cent magnesium have a high mechanical strength and, although their corrosion resistance is lower than the 5 per cent alloy, they may be regarded as high-strength casting alloys with fair corrosion stability. These materials were primarily developed for marine applications and in sea conditions have undoubted advantages over the silicon alloys for the preparation of important components. However their rapid oxidation at high temperatures and poor casting properties limit their use to castings of simple geometrical shape.

Aluminium-Silicon Alloys Casting alloys containing 5 to 14 per cent silicon are widely used. Aluminium and silicon form a simple eutectic system with an eutectic concentration of silicon of about 11. 6 per cent. The solubility limit of silicon at the eutectic temperature is 1. 3 per cent but at room temperature is less than 0. 1 per cent. These alloys have very good casting properties, particularly at the eutectic composition, and their corrosion resistance in sea water is satisfactory, particularly when the silicon content is low. Their stability largely depends on the non-separation of intermetallic compounds and the retention of the eutectic in a thinly dispersed state.

BERYLLIUM

The main uses of beryllium are in nuclear reactor environments and as a minor alloying element, e.g. in the beryllium bronzes and nickel-beryllium alloys. As an alloying addition it improves both the mechanical properties and the corrosion resistance of the parent metal. The addition of 2. 0-2. 5 per cent to copper increases the hardness, tensile and fatigue strength, and also its resistance to wet corrosion.

In chemical behaviour beryllium is similar to aluminium and magnesium. Like aluminium, zirconium and titanium its resistance to corrosion is due to the formation of a protective oxide film on its surface. It is an amphoteric metal and is attacked by both acids and alkalis but in near-neutral waters it generally has a good corrosion resistance.

In dilute hydrogen peroxide at 85°C it is sensitive to copper ions at a concentration of less than 1 ppm and to a less extent to iron ions. In one year the rate of attack in 0. 0005 molar hydrogen peroxide at below 100°C was 0. 13-0. 26 mdd on the average but this attack was always accompanied by pitting. The annealed material is less resistant than machined or pickled surfaces. Galvanic effects can occur when beryllium is coupled to stainless steel.

In distilled water at 270°C the corrosion rate varies between 0. 77 and 27 mdd with pronounced pitting. Above 260°C the corrosion resistance becomes progressively poorer and protective cladding may be necessary.

Aeration and movement are beneficial and at a flow of 5-8 ft/sec at 85°C the frequency of pitting is reduced. In the same flow range at 80°C the corrosion rate in the pH range 4-8 is 0. 13-0. 39 with a maximum pit depth of 3. 5 mils.

Dissolved chlorides increase the corrosion rate. At a chloride ion concentration of 0. 5 ppm widespread pitting occurs, at 1 ppm the rate of attack is 1. 7 mdd while at 10 ppm the rate is 17 mdd.

CADMIUM

The potential of cadmium is about —0. 52 volts in chloride solutions and it has only a very weak tendency towards passivation. It is comparable to zinc in many properties including its corrosion behaviour but, owing to its less negative potential, it is more stable in acid and neutral solutions and, unlike zinc, is stable in alkalis.

Cadmium is used mainly for the provision of anodic coatings on steel and to some extent on aluminium alloys. In the plating of steel articles cadmium has an advantage over zinc; due to its more positive deposition potential less hydrogen is usually evolved and hence there is a reduction in hydrogen embrittlement. Ferrous metals coated with cadmium are used in the manufacture of laundry and dish-washing equipment as well as equipment used under conditions where steel has a tendency to failure by corrosion fatigue.

Owing to its less negative potential cadmium is not capable of protecting steel to the same extent as zinc at breaks in the coating. Furthermore, like zinc, there is a tendency to potential reversal which can cause pinhole pitting of iron at pores in the coating.

Alkali metal carbonates dissolve cadmium much more rapidly than chlorides or sulphates. In quiescent and aerated sea water the attack on cadmium is less than 1.8 mdd, much lower than that on zinc.

COBALT

Pure cobalt is of little practical use and is mainly used in alloys. Wear resistant alloys based on cobalt are not corroded to any appreciable extent by mine water, sea water or mains water at ordinary temperatures. In sea-water tests over a period of 2 years these alloys gave an average corrosion rate of 0.6 mdd with a maximum pit depth of 7 mils. The attack on electrolytic cobalt is 4.2 mdd.

Impingement tests on a 65/30/5 cobalt-chromium-tungsten alloys have shown that its behaviour is comparable to a 12.0 per cent chromium steel.

CHROMIUM

Chromium and its alloys are rarely used industrially since chromium is very heavy and normally brittle although very pure chromium is ductile. The brittleness is caused by its very large affinity for oxygen, nitrogen and carbon; when melted in air it contains up to 0.2 per cent nitrogen. The purest electrolytic chromium may contain up to 1.5 per cent of chromium oxides. Consequently it is mainly used for the preparation of alloys with high corrosion resistance, such as stainless steels or nichrome; as a coating on other metals to provide a surface which is hard and resistant to wear, erosion and corrosion, or merely for decorative purposes.

In decorative applications very thin coatings of chromium, 0.01 to 0.06 mil, are applied over a suitable thickness of nickel or a composite coating of copper and nickel. Thicker coatings, varying generally from 0.1 to 10.0 mils, are plated directly in engineering applications. In salvage operations, on such articles as crankshafts and diesel-engine liners, where there has been appreciable loss in dimension due to wear, improper machining or corrosion, deposits up to 30 mils in thickness may be required.

Chromium owes its high resistance to corrosion to its extraordinary
ability to transform into the passive state under the influence of both oxidants and dissolved oxygen and it passivates spontaneously under natural
conditions. This condition is usually associated with the formation of a
thin dark-green layer of chromic oxide, Cr_2O_3, which is quite refractory
and chemically stable, being insoluble in water and acids.

The protection afforded to the basis metal can range from poor to excellent depending mainly on the thickness of the coating. Strong galvanic
couples can develop between the cathodic coating and a less noble basis
metal such as steel, copper, zinc and aluminium, and for complete protection the coating must be sufficiently thick to be without porosity. With
thin coatings from 0.01 to 0.02 mils an undercoating of nickel is necessary owing to the inability of the chromium deposit to remain free from
cracks and discontinuities. The thickness has to be increased to about
0.5 to 1.0 mil before the porosity is reduced to a level at which the protection of the basis metal is satisfactory. Above 1.0 mil a high degree of
protection from corrosion is afforded by chromium coatings.

Chromium is not corroded in most salt solutions in the presence of oxygen from the air. It is not stable, however, in reducing media or in solutions containing appreciable amounts of chloride.

COPPER AND ITS ALLOYS

In spite of its scarcity and its mechanical strength being inferior to that
of many of its alloys, <u>pure copper</u> has many technological uses because of
its high chemical stability in a number of corrosive media and its high
heat and electrical conductivity. The range of temperature over which it
can be used is, however, limited owing to its decreased strength and the
formation of relatively non-protective oxides at high temperatures.

COPPER

The corrosion resistance of copper is due to its being a relatively noble
metal. Its satisfactory service in waters depends on the formation of relatively thin adherent films of corrosion products, e.g. cuprous oxide and
basic copper carbonate. It has only a weak tendency to passivation and
hence the effect of unequal aeration is very slight. However the influence
of copper ion concentration on the potential of copper in solution is very
marked. For this reason when there are varying solution velocities over
a copper surface, e.g. when the solution is stirred, the parts washed by
solution with the higher rate of movement become anodes and not cathodes
as would be the case with iron (Chap. I).

In a solution containing complexants such as ammonia or cyanide, particularly with access of oxygen from the air or in the presence of oxidants,
copper corrodes rapidly with the formation of complex ions. This fact
shows that in the corrosion of electropositive metals such as copper the
ease with which the anodic process can occur is all important. When the
anodic process is possible, as in the presence of complexing agents, high

rates of attack only occur when a strong cathodic depolariser such as oxygen or other oxidant is present.

Distilled water is not very corrosive to copper but over long periods the water will pick up traces of copper. Hence in tanks for holding distilled water or for water storage on ships the copper should be tinned.

In soft waters, particularly those containing appreciable amounts of free carbon dioxide, and in carbonated waters in general, the corrosion of copper is much greater. Thus while the initial corrosion rate in distilled water may be 12 to 37 mdd the rate in an aggressive supply water may be as high as 62 mdd. Such waters may be corrosive enough to pick up enough copper to form green stains on plumbing fixtures.

Hard waters are seldom corrosive to copper because of the protective film of calcium compounds which soon forms. When waters with an appreciable temporary hardness are softened, however, they can become corrosive, especially if heated above 140°C. The calcium compounds necessary to form a protective film are then absent and the sodium bicarbonate formed breaks down with the release of carbon dioxide.

Copper is very stable in salt solutions and sea water. When constantly immersed the corrosion rate of pure copper in sea water is 5 to 17 mdd and under varying immersion, as at half tide, the rate is 5 to 25 mdd. In the latter case arsenical copper is rather more resistant than tough-pitch copper. Under constant immersion the corrosion resistance of copper is 2 to 5 times greater than that of mild steel and under half-tide immersion the advantage of copper is even more notable. Copper is satisfactory in stagnant sea water or where the velocity is less than about 3 ft/sec. The rate of dissolution of copper is, however, large enough to render it toxic to marine boring animals, hence its value as a sheathing material for wooden craft and piling. For the same reason copper compounds are used in marine anti-fouling paints.

Copper is not sufficiently resistant in rapidly flowing sea water to be suitable for condenser tubes in ocean-going ships or in power stations using tidal water. A number of copper alloys have superior resistance under such conditions (Chap. VIII).

Thin carbonaceous films have resulted in pitting of copper tubes. These are derived from carbonisation of the lubricant used in manufacture and by tightening up the procedure these troubles have now been largely eliminated. Films of manganese oxides, derived by slow deposition from soft moorland waters, can also give rise to troubles with localised attack.

A beneficial effect, presumably due to the presence of an organic chemical at very low concentration, is observed in the corrosion of copper in certain waters. The organic agent responsible for this action has not been isolated and the way in which it operates is imperfectly understood.

BRASSES

The common brasses are alloys of copper with 10-50 per cent zinc and often a number of other components including tin, iron, manganese, aluminium and lead. Zinc dissolves in copper up to 39 per cent to give a single phase alloy, α-brass, and with zinc contents in the range 47-50 per cent the alloy is again single phase, β-brass. In between these two, i.e. with 39-47 per cent zinc, the alloy contains both phases, $\alpha + \beta$ brass. In the corrosion of the α-brasses the constituents are dissolved according to the copper-zinc ratio but as the two-phase alloy is approached appreciably more zinc starts to go into solution. In the duplex brasses dezincification takes place by the favoured dissolution of the less noble β phase and in the pure β alloys only zinc appears in the corrosion products, the copper being redeposited as metal.

The stability of brasses in natural waters depends in a complex manner on the dissolved salts, the hardness, the dissolved gases and on the formation of protective films. In general the brasses have a high stability with rates of attack varying from 0.6 to 6 mdd. The attack is increased by higher concentrations of carbon dioxide and with the higher zinc brasses is accompanied by dezincification. Tin additions are effective in reducing dezincification and red brass, 85/15 copper-zinc, or naval brass, 63/36/1 copper-zinc-tin, are commonly used alloys. For fresh-water plumbing, red brass piping is preferred to copper alloys such as Muntz metal, 60/40 copper-zinc, and leaded yellow brass, 67/33 copper-zinc, which have a tendency to dezincification in some waters.

The attack on brass by pure water such as condensate is very small, less than 3 mdd, but this is appreciably increased in the presence of air, carbon dioxide or ammonia. The scale deposited from hard waters becomes less protective at elevated temperatures owing to the decrease in adhesion.

In sea water the rate of attack is generally small, between 2 and 24 mdd, but may be much greater since it varies with composition and local conditions. Brass with a copper content of about 70 per cent is the most stable of the straight brasses in sea water, and Admiralty brass, 70/29/1 copper-zinc-tin is often used. If the copper content is higher than this there is a tendency to local attack, particularly at the water line, and if the copper content is lower there is an increased likelihood of dezincification. The high tensile brasses with a higher zinc content are more resistant to corrosion-erosion and cavitation and are used in the manufacture of propeller screws. Their inclination towards dezincification is reduced by small additions of arsenic, antimony or phosphorus. The corrosion stability of brass is considerably increased by addition of aluminium and the brass 76/22/2 copper-zinc-aluminium is widely used in marine applications, e.g. in condenser tubes.

Acid mine waters discharged into rivers make the natural water more corrosive owing to the presence of small amounts of sulphuric acid and ferric sulphate. Copper alloys containing tin are the most resistant to this type of water.

BRONZES

In the past the term 'bronze' always referred to the copper-tin series of alloys but now it is generally applied to casting alloys based on copper

irrespective of whether tin is present or not. In this book the term bronze
is used to indicate the copper-tin series of alloys; other bronzes are
qualified accordingly e.g. aluminium-bronze. A further complication is the
common use of the term 'manganese-bronze' to dsignate an alloy which is
essentially a 60/40 high tensile brass. This material is referred to as
manganese-brass.

Copper can form a solid solution with tin containing up to 15.8 per cent
tin, the α-phase. The actual solubility at room temperature is lower but
the decrease in concentration with falling temperature is so very slow
that the solubility below 520°C for normal alloys can be taken as 15.8 per
cent. The bronzes containing up to about 8 per cent tin are used in the
rolled form and with higher percentages in the wrought form. Further
alloying additions of small amounts of phosphorus, zinc and lead are made.
Additions of iron, antimony and bismuth are dangerous and are tolerated
only up to 0.2 to 0.5 per cent.

The attack by water depends on the oxygen, carbon doxide and salt content
and the possibilities of the formation of protective layers. Both rolled
and wrought bronzes with and without additions of lead or zinc may be
used. Bronzes are very stable against steam but under more severe
corrosive conditions and high steaming rates attack up to 200 mdd can
occur. They are seldom used above 260°C and 100 atm. pressure.

In sea water the corrosion rate of bronzes varies from 3 to 8 mdd
according to local conditions. In shipbuilding, bronzes with more than 5
per cent tin and also additions of lead and zinc are used. These include
the well-known Admiralty bronze or Admiralty gun-metal, 88/10/2 copper-
tin-zinc, and naval bronze, 88/8/4 copper-tin-zinc. An important function
of the zinc is as a deoxidiser to prevent the formation of undesirable tin
oxide inclusions. In certain cases the attack on lead-tin bronzes with tin
contents greater than 10 per cent is increased to 22 mdd.

Aluminium Bronzes These usually contain not more than 9-10 per cent
aluminium and sometimes small additions of manganese and nickel. In sea
water they have a higher stability than the other copper alloys with corrosion rates only one-tenth of that of bronze and one-thirtieth of that of
brass. The rate of attack on an 8.0 per cent aluminium alloy is 0.3-
0.8 mdd at 30°C and 2.3 mdd at 60°C. More complex alloys containing up
to 11.5 per cent aluminium, 5.5 per cent nickel, 5.0 per cent iron and
3.5 per cent manganese are casting alloys used in the manufacture of
ships' screws. These alloys are able to withstand the severe demands of
service in Arctic waters better than manganese brasses and are much
more stable towards erosion and cavitation.

Silicon Bronzes These contain up to 4.5 per cent silicon and sometimes a
number of other additions. The 3.0 per cent silicon bronzes are stable in
natural waters, including sea water, and are very suitable for hot-water
equipment and for screws used in marine fittings.

COPPER-NICKEL ALLOYS

Alloys based on copper and containing from 5 to 40 per cent nickel have good mechanical properties at moderately elevated temperatures and excellent resistance to corrosion in many environments and in particular in contact with brackish water and sea water. They are more stable than the brasses in flowing water, less susceptible to stress corrosion, and are widely used in shipbuilding for installations involving heat exchange.

In this group of alloys most attention has been directed to 70/30 copper-nickel and 90/10 copper-nickel and the improvement in corrosion resistance under flow conditions obtained by small additions of iron and manganese. In both alloys the optimum performance, even at flow rates up to 16 ft/sec, is obtained with an iron concentration of 1 per cent, viz. 69/30/1 and 89/10/1 copper-nickel-iron. These alloys are to be recommended for brackish waters, sea waters and waters with total dissolved solids of more than 2000 ppm, i.e. 0.2 per cent. They have an advantage in that they do not require cathodic protection as is provided for aluminium brass tubes, 76/22/2 copper-zinc-aluminium, by iron anodes bolted to the water boxes or by the boxes themselves.

69/30/1 copper-nickel-iron and also alloys of the Monel type, e.g. 30/66 copper-nickel with 4 per cent iron and manganese, are used for high-pressure feed-water preheaters. The introduction of copper ions into the boiler, which might be dangerous, is thereby avoided. In recent years, troubles have arisen during the use of 69/30/1 and 79/20/1 copper-nickel-iron alloys for tubes; thick layers of corrosion product form and subsequently flake off. This type of attack is called <u>exfoliation</u> and is limited to heaters which are operated discontinuously and which are used with feed waters treated with sodium sulphite. It may be avoided by using the 89/10/1 copper-nickel-iron alloy which can also be used for equipment evaporating sea water to provide drinking water on board ship.

The 90/10 copper-nickel alloy and the silicon bronzes have very good mechanical properties and are readily welded. An important use is in the fabrication of storage tanks for hot fresh water.

The 69/30/1 copper-nickel-iron is used for dealing with condensate and feed water and has a corrosion rate of less than 19 mdd. At 70°C under air containing carbon dioxide the attack is higher, 64-106 mdd at 16 atm pressure. This alloy and the corresponding one containing small amounts of manganese are resistant to cavitation attack.

LEAD

The corrosion resistance of lead is largely determined by the solubility of the lead compounds formed and their physical properties. Where the products of attack are soluble, as in nitrate solutions, corrosion can proceed unhindered but insoluble products, such as those formed in sulphate solutions, can provide highly protective films on the metal surface. These protective films can form over a wide range of pH values from about 3 to 11 which includes all natural and treated waters. Where the products form a continuous and adherent layer the rate of attack under the best conditions is less than 0.8 mdd.

The metal oxide is not very protective and aerated distilled water free from dissolved carbon dioxide is moderately corrosive. In soft waters, such as distilled or rain water, the corrosion rate is proportional to the concentration of dissolved oxygen and is increased by agitation or aeration.

Two reactions can occur depending on the carbon dioxide content. If this is high, a basic carbonate, and possibly some oxide, is formed on the surface and the rate of attack falls off, the reaction then being under anodic control. If, on the other hand, the carbon dioxide concentration is low the basic lead carbonate is precipitated in the solution as a white cloudiness and attack proceeds at a uniform rate. The carbon dioxide does not reach the surface of the lead in sufficient amount to form a protective layer but is previously removed from solution by reaction with dissolved lead dioxide. In this case the corrosion reaction is under cathodic control.

Lead also dissolves in underground waters containing organic acids and in acid mine waters. Small amounts of nitrate increase the corrosive attack on lead since lead nitrate is very soluble.

Soft waters generally attack lead sufficiently to build up a lead concentration of more than 0.1 ppm and since this is physiologically dangerous, lead is no longer used for conveying potable soft waters. Lead is however stable in soft waters containing calcium sulphate, calcium carbonate or silicic acid. If the water is not too soft, both carbonate and sulphate can share in the formation of a protective layer.

The corrosion rate in sea water is only 4 mdd, i.e. less than that of iron, and lead is, therefore, used with sea water in the form of sheet and piping and as a protective coating on copper and other metals. Lead-tin alloy can be applied to copper and steel by hot dipping, electrodeposition or wiping. On copper it is used extensively in ships piping carrying sea water. 'Terne-plate' is steel which has been coated with a lead-tin alloy with a lead content of 75 to 90 per cent depending on its use.

In harbour water 99.5 per cent sheet lead gave uniform attack with no pitting and an average corrosion rate of 1.6 mdd.

Owing to its softness when pure, lead is usually used with various alloying additions. Antimonial lead contains up to 12 per cent antimony, but more usually 4-6 per cent, and may be used up to 120°C. Tellurium lead containing 0.05 per cent tellurium is a work-hardening alloy with a high resistance to fatigue failure. Soft solders are lead-tin alloys of varying composition and will be discussed later.

MAGNESIUM

Magnesium is the lightest of all the constructional metals with a specific gravity of 1.74 but its use is limited by its high chemical reactivity and low corrosion resistance. It is rarely used in the unalloyed state and although up to 10 per cent aluminium, 3 per cent zinc and 2.5 per cent manganese are added to improve its mechanical and other properties, the corrosion resistance of these alloys is generally inferior to that of pure magnesium.

In most environments magnesium and its alloys corrode with the hydrogen-evolution type of attack and only in media such as distilled water does appreciable oxygen reduction occur. Active ions such as chloride increase the rate of the corrosion process on magnesium. The corrosion stability of magnesium alloy castings can be markedly reduced by the use of fluxes containing chloride. As a consequence of the similarity in their specific gravities the flux is able to penetrate with ease into the molten alloy.

In corrosion by salt solutions, magnesium and its alloys are affected to a larger extent than aluminium by impurities in the alloy and by contact with other metals. The corrosion potential of magnesium is more negative than any other structural metal or alloy and hence in any macro-cell set up by contact with such alloys magnesium is invariably the anode.

The corrosion resistance of pure magnesium is reduced by very small amounts of more noble metals such as iron, nickel, cobalt and copper present as impurities. Such metals have a low hydrogen overvoltage, i.e. hydrogen is readily evolved from their surface. In chloride solutions the tolerance limits of impurity metals are 0.017 per cent iron, less than 0.0005 per cent nickel and 0.1 per cent copper. Metals with a higher hydrogen overvoltage, i.e. on which hydrogen evolution is more difficult such as lead, zinc and cadmium and even metals with a more electropositive potential such as manganese and aluminium, affect the corrosion behaviour of magnesium to a less serious extent.

The tolerance limit for iron is further reduced by the presence of aluminium in the magnesium alloy. A few hundredths of one per cent of aluminium reduces the tolerable concentration of iron to a few thousandths of one per cent. The addition of one per cent zinc or manganese reduces the high corrosion rate that would otherwise result if this limit were exceeded. Manganese forms an alloy with the iron-rich inclusions and thereby increases the hydrogen overvoltage.

Intense localised attack may usually be traced to the presence of particles cathodic to magnesium either embedded in the surface or dispersed in the metal. In order to minimise this pitting tendency steps must be taken to remove heavy metal impurities from the wrought alloys. Contamination of the surface by iron invariably occurs on rolling magnesium alloy sheet. Even if the rolls are specially treated to prevent pick-up and the sheet is acid pickled, traces of iron sometimes remain strongly embedded in the sheet.

Cast alloys have a much greater susceptibility to corrosion and pitting. Removal of adherent foundry sand from castings by blasting with shot or grit, or by emerying, lowers the natural corrosion resistance of the surface and also its ability to form a protective film. Particles of metal or emery are picked up by the surface and act as efficient cathodes. Glass or carborundum papers are usually relatively harmless.

Many commercial abrasives reduce the corrosion resistance to such a marked extent that a piece of metal that would otherwise remain unattacked in salt water for several hours will, after shot blasting, evolve hydrogen vigorously and the rate of attack will be increased several hundred times. Little improvement is gained by acid pickling but anodising at high voltage in a solution of ammonium fluoride is very effective in removing foreign metal particles.

Graphitic carbon derived from lubricants used in fabrication can also form an active cathode. In the above anodising treatment this is undercut and insulated from the basis metal by a layer of magnesium fluoride. It is then less harmful and is more easily removed by subsequent treatment in chromic acid or hot caustic soda solution.

In relatively pure waters and distilled water at ordinary temperatures, magnesium alloys rapidly form a protective film of magnesium hydroxide which restricts further action. If the water contains small amounts of chloride, heavy metal salts or carbon dioxide, pitting may take place at points where the protective film has broken down and if the protective film is destroyed or prevented from forming, e.g. by agitation, more widespread corrosion can occur. In small volumes of stagnant water the corrosion of magnesium is negligible, but it increases if the water is constantly replenished since the water never becomes saturated with magnesium hydroxide.

Nitrates or phosphates do not affect the corrosion of magnesium and its alloys. Corrosion is appreciably reduced, even in solutions containing chlorides, if the water is made alkaline since magnesium is stable in ammonia and alkali solutions at temperatures up to 50-60°C.

The lower rate of attack in sea water compared with that in a solution of the same chloride concentration, i.e. 3 per cent sodium chloride, is attributed to the presence of sulphate. In 3 per cent sodium chloride solution the rate of attack is increased by adding magnesium chloride but reduced by adding the sulphates of magnesium, calcium or potassium.

Magnesium alloys are susceptible to intercrystalline corrosion and corrosion cracking particularly when they are deformed and in a stressed state. Corrosion cracking under stress is likely under conditions where general corrosion is localised by the alloy being partially passivated, e.g. in distilled water containing small concentrations of chloride. Tensile stresses have little influence on general corrosion but often produce corrosion cracking of higher strength alloys such as the alloy containing 3.5 per cent aluminium and small amounts of zinc and manganese. This attack is characterised by the transcrystalline nature of the cracks.

Cast magnesium alloys, mainly those containing 6 per cent aluminium and 3 per cent zinc or 8 per cent aluminium and 0.5 per cent zinc together with small amounts of manganese, are widely used in the cathodic protection of steel structures in the sea or in the ground. The more negative potential of magnesium alloys makes them more efficient than alloys based on zinc and aluminium. In spite of the appreciable self-dissolution of magnesium, the efficiency of a magnesium protector, i.e. the maximum amount of electricity per unit weight of dissolved material, may be greater than zinc because of its smaller electrochemical equivalent.

MANGANESE

The amount of manganese in the earth's crust is greater than that of such widely used metals as copper and zinc. It has not yet been found possible to prepare it in a sufficiently pure state to be fabricated and even when distilled *in vacuo* it is very hard and brittle. If these difficulties could be surmounted it might find use as protective coating for steel since, like zinc and cadmium, it has a very negative potential.

When cold, water acts very slowly on manganese but on heating the attack is much more rapid.

MOLYBDENUM

Technically molybdenum is a very interesting and important metal but, owing to its scarcity, it has a limited use as a constructional metal. It is however widely used as an alloying addition to steels.

Cold and boiling water do not attack molybdenum and it is stable in most salt solutions, including chlorides, and in sea water.

NICKEL AND ITS ALLOYS

Although in mechanical properties nickel is similar to mild steel its corrosion resistance is relatively very high. It is more electropositive than copper, chromium, iron, aluminium and zinc. Nickel differs from copper in having an appreciable tendency to transform to the passive state and to a large extent owes its high corrosion resistance to an ability to develop a resistant oxide film on its surface.

Commercial wrought forms of nickel usually contain at least 99 per cent nickel, most of the balance being made up by manganese and iron. Depending on the refining process it may also contain small amounts of cobalt which is neither harmful nor helpful. High purity nickel has limited uses as a constructional material and its main application is where high stability to alkalis is required. Important uses of nickel are as an alloying component in stainless steels and other alloys or as a protective and decorative coating.

Nickel coatings are mainly applied by electroplating and coated steel may be fabricated into containers by welding.

The use of alloys based on nickel is limited by their cost although they are very resistant to attack by most natural and treated waters. In both flowing and quiescent waters the corrosion rate is negligible, no increase in attack is produced by aeration and agitation is more of an advantage than a detriment. A special 85/15 nickel-copper alloy has been used for hot water heaters as well as for critical parts of valves and water meters. Some of these alloys are also used in high-purity waters at high temperatures.

Both ordinary nickel and those of its alloys which age harden are highly resistant to impingement attack and erosion and give good performance in, for example, sea water at high velocity. In sea water which is only moving slowly or is stagnant both nickel and its alloys with copper are susceptible to severe local attack. Pitting takes place beneath deposits, fouling organisms or foreign bodies which adhere to the surface and form oxygen concentrations cells.

Nickel can be safely used for the handling and storage of high purity water where it is desirable to minimise contamination. The corrosion rate in distilled water is less than 0.012 mdd. Inconel, 77/16/7 nickel-chromium-iron, is preferred to austenitic stainless steels in high-temperature water since it is resistant to stress corrosion cracking from trace amounts of chlorides.

Although under normal conditions nickel is stable towards condensate, severe corrosion can occur in hot steam condensate under pressure when it contains appreciable amounts of oxygen and carbon dioxide. Nickel-copper alloys show a similar sensitivity although Inconel is more stable under these conditions. When nickel is used for handling steam condensate, as in evaporators or heat exchangers, the boiler feed-water should be deaerated or as much as possible of the non-condensable gases vented from the operating unit.

Nickel has a high resistance to attack by most fresh waters and in river, lake and well waters its corrosion rate is normally less than 0.6 mdd. Dissolved hydrogen sulphide and carbon dioxide do not increase the attack significantly although hydrogen sulphide produces some discolouration of the metal.

In domestic hot water the corrosion rate of nickel is normally less than 0.12 mdd at temperatures up to 95°C although in a few locations the attack may be as high as 1.2 mdd. Nickel and nickel-clad alloys are used for hot-water storage tanks.

Appreciable corrosion is experienced in acid mine waters containing significant amounts of ferric and cupric salts. In general where the amount of industrial contamination is high increased corrosion rates may be expected.

NICKEL-COPPER ALLOYS

Nickel and copper form a continuous series of solid solutions. The copper-rich alloys have corrosion characteristics close to copper and have already been discussed. When nickel is the major phase their corrosion behaviour is closer to that of nickel. Of the alloys of the nickel-copper type, containing up to 40 per cent copper, one of the most used is monel, which contains about 30 per cent copper and 3-4 per cent iron and manganese, and sometimes aluminium and silicon. This material has a higher corrosion resistance in salt solutions than either pure copper or nickel, and owing to its stability in sea water is used for coolers, condensers, sea water evaporators, and in the form of castings, for ships' screws, pumps, valve seatings, nails and so on.

The hardenable alloy monel-K is used for parts of machines bearing appreciable loads, e.g. parts of centrifugal pumps and shafts in contact with sea water. It is also used for bolts where steel and other materials are insufficiently resistant to corrosion or may undergo stress corrosion.

In sea water Inconel, a nickel-chromium alloy, undergoes local corrosion in stagnant regions and the most stable alloys under these conditions are those of the Hastelloy type, i.e. nickel with chromium or with molybdenum and chromium, which are not susceptible to localised attack.

Nickel and many of its alloys have a lower resistance than austenitic steels to attack by oxygenated waters at 260°C; the attack on nickel and Inconel is 0.3-1.0 mdd, on monel 1.0-1.8 mdd and on Invar, 64/36 iron-nickel, 3 mdd. In oxygen-free water or water saturated with hydrogen, e.g. in the primary circuit of a nuclear reactor, these alloys are stable and even at flow velocities of 30 ft/sec the rate of attack is less than 0.17 mdd.

NIOBIUM (COLUMBIUM)

Niobium is a metal of high corrosion resistance owing to the readiness with which it forms protective oxide films. By anodic treatment oxide films of very high corrosion stability can be formed. Niobium is unattacked by fresh water or sea water.

At the present time its main application is as an alloying element in stainless steels. The carbon is sequestered as niobium carbide and the formation of chromium carbide during fabrication by welding is prevented. It is likely that with increase in the extraction of niobium its applications may grow.

TIN

Owing to its scarcity, which makes its cost relatively high, and its poor mechanical properties the use of pure tin is very limited. In aqueous enviroments its use is restricted to pipes for handling very pure water.

The main applications of tin are as protective coatings or liners for stronger but less corrosion resistant metals and as an alloying element in bronzes or solder. As a protective coating it is mainly applied on iron—and the uses of tinplate are continuously increasing—and on copper and brass. In the form of sheet it is used for lining tanks and pipes where the purity of the water is of primary importance. Pipes of tin may be reinforced by inserting in a brass pipe and joining the two by special techniques.

Tin coatings may be applied by hot dipping or by electrolytic methods. Coatings prepared by machine are generally 0.04-0.2 mil thick but thicker coatings, up to 1 mil, may be obtained by manual application or wiping.

Tin is cathodic to iron but the corrosion behaviour of tin-plated steel is governed by a number of factors including the purity of the tin, the porosity and thickness of coating, the method of tinning and the type of steel. Careful attention should be paid, however, to the possibility of galvanic cells arising between the plating and the basis metal, particularly in soft waters.

In strong chloride solutions tin has the relatively noble corrosion potential of —0.25 volts and only a weak tendency to passivate. Its good behaviour in near-neutral solution is indicative of its corrosion resistance in all fresh waters whether hot or cold. In soft and distilled water, surface films are formed and it has an acceptable stability in aggressive waters containing dissolved carbon dioxide or chlorides and sulphates.

In hard waters at or above ambient temperature the increased attack in the region of the water line beneath deposited carbonate may be eliminated by alloying with 5 per cent antimony. Such alloying additions also increase the hardness of the material.

In sea water the pitting behaviour of pure tin may be inferior to commercial grades. Thus it has been shown that the average rate of penetration of pure tin was three times the figure of 0.15 mdd obtained with tin containing 0.3 per cent antimony and 0.4 per cent lead.

SOLDER

Solders comprise a wide range of alloys of tin with lead or zinc. Many troubles arising from soldered joints can be attibuted to the action of flux residues hence the choice of a suitable flux is very important.

Although tin-lead solders are anodic to copper this is usually of little practical significance. However in waters of high conductivity, such as sea water, corrosion products are formed on these alloys which are cathodic to copper and when used as coatings this can give rise to severe corrosion at bare spots.

Tin-lead solders are cathodic to aluminium and tin-zinc alloys give joints with a higher resistance to corrosion than the lead alloys. With all the tin-base alloys the vulnerable zone is the diffusion layer between solder and aluminium. This layer is less noble than aluminium, tin or lead but more noble than zinc so that corrosive attack on it is less rapid with the zinc alloy than with the lead alloy. The eutectic alloy, 91/9 tin-zinc, is the best low-temperature solder for aluminium; it is easy to apply and has better resistance to corrosion.

TITANIUM

Of the newer metals titanium has already found an important place among structural materials. Owing to the technological difficulties of producing titanium in a sufficiently pure state its cost is still quite high and its availability is limited. For this reason the use of titanium is restricted on economic grounds to very aggressive conditions such as exist in pumps dealing with brackish water and where a high corrosion resistance is very necessary as in atomic reactors. Platinised-titanium anodes are used as economical non-consumable electrodes for the cathodic protection of ships jetties and pipelines.

In aerated chloride solutions at 25°C titanium has a corrosion potential of about +0.18 volts. Only the noble metals and several metals in the passive state, e.g. stainless steels, have a comparable potential in these solutions.

The high corrosion resistance of titanium is due to the stability of the passive state. Unlike iron, nickel, chromium, and the stainless steels, titanium retains its passive state not only in solutions containing oxygen but also in the presence of chloride ions at practically any concentration and this resistance to neutral and weakly acid chlorides is characteristic.

Titanium is resistant to stress corrosion cracking in chloride solutions and even under conditions severe enough to produce failure of 18/8 stainless steel in a few hours.

Titanium has very high resistance to corrosion by sea water and is more resistant than materials such as stainless steel, monel and cupronickel and even approaches the resistance of the noble metals. In both stagnant and moving sea water titanium is not susceptible to pitting and crevice attack and furthermore resists cavitation, corrosion fatigue and stress corrosion.

The corrosion of titanium is not affected by contact with other metals but will increase the corrosion of most metals, even copper. At contacts it behaves in a similar way to stainless steel; the cathodic reduction of oxygen does not readily take place on the passive surface, i.e. the reaction is very easily polarised.

TUNGSTEN

The main use of tungsten is as an alloying element. In chloride solution its corrosion potential is close to zero and in chemical and corrosion properties it is similar to molybdenum.

Tungsten is not attacked by hot or cold water.

URANIUM

The most important application of uranium is in the nuclear energy field. It is not attacked by water at ordinary temperature but the rate in boiling water is over 7000 mdd. At 300°C the rate of attack is 2000 times that at the boiling point.

Below 90°C, oxygen-saturated water produces extensive pitting but above this temperature general attack occurs irrespective of the oxygen concentration. Displacement of the oxygen by hydrogen or inert gases produces general attack at temperatures much lower than 90°C. The corrosion rate in hydrogen saturated water is 30 times greater than the initial rate in aerated solutions indicating that the oxygen forms a protective film.

Alloys containing 5-6 per cent zirconium after adequate heat treatment are able to develop good protective films in boiling distilled water and corrode at a rate considerably less than that of the unalloyed uranium.

ZINC

Although the corrosion potential of zinc, about —0.86 volts in strong chloride solutions, is more negative than all other metals with the exception of magnesium, it has a wide field of use in natural waters owing to its ability to form protective layers on its surface. Its corrosion rate is only a fraction of that of steel, hence it has an important application as a coating. It also has the secondary advantage of being able to protect the basis metal cathodically at adventitious breaks in the coating. Zinc coatings are less costly and more protective than other coatings and are preferred to nickel, tin, copper and lead which tend to accelerate attack on exposed steel.

Paints increase the lifetime of the coating and galvanising is an effective pretreatment for painting by reducing rusting and costly derusting work in later paint maintenance. Zinc-rich paints, containing not less than 95 per cent zinc, act in a similar manner and are very valuable as primers in patching up painted galvanised structures.

The normal impurities in zinc are lead, cadmium and iron. A small iron content increases the resistance to attack owing to an increase in the protective character of the corrosion product. This is, of course, undesirable in zinc used to make anodes for cathodic protection (see Chap. XI). Copper is sometimes added to zinc in order to increase the hardness of rolled products. Additions of 1.5-2.5 per cent copper and 0.5-4.5 per cent aluminium improve the casting properties of zinc but such alloys have a low corrosion resistance in water and water vapour.

Zinc does not have any appreciable inclination to passivation except in chromate solutions in which a protective film of zinc chromate is formed. In neutral solutions the main cathodic reaction in the corrosion of zinc is the reduction of oxygen and alloying additions do not markedly affect the resistance to attack.

Zinc is amphoteric and corrodes in both acid and alkaline solution. It is most stable in the pH range 9-11 and the greater the deviation in pH from these limits the higher the corrosion rate. The use of zinc is not practicable if the pH of the water is much less than 7 or more than 12 since the protective hydrate film dissolves and corrosion increases.

The behaviour of zinc in natural waters is largely conditioned by the type of dissolved salts and gases. In the presence of carbon dioxide and calcium and magnesium salts protective layers of basic carbonate or mixtures with calcium and magnesium carbonate are formed. Consequently the dissolution rate decreases with time in natural and distilled water. In stagnant solution of limited oxygen content, zinc is susceptible to attack owing to the formation of concentration cells.

The rate of attack in distilled water is from 20-22 mdd falling to about half this value over long periods of time. Owing to the small solubility of zinc hydroxide protective layers may form by interaction with dissolved carbon dioxide. In this, zinc differs from lead in distilled water where the formation of carbonate may take place in the solution and not on the metal surface.

The rate of attack in fresh waters varies widely but 7-15 mdd is a fairly typical range for aerated quiescent waters at ordinary temperatures particularly where there is scale deposition. At 70°C the rate may increase up to 50 mdd.

Distilled water is more corrosive than waters containing dissolved salts. In stagnant distilled water the attack is 4 to 6 times as great as in sea water. In flowing sea water rates of attack vary from 0.5-9.0 mdd where the protective coating is mainly carbonate, hydroxide and oxychloride.

Although zinc and zinc alloys are rarely used on their own in sea water, zinc is however quite often used as a protective coating. At small rates of movement a zinc coating will decrease in thickness by approximately 1.2 mils per year. The decision that has to be made, therefore, is whether it is more economic to lose 1.2 mils per year of zinc coating or 4.8 mils per year of steel on account of the corrosion of the unprotected basis metal. From these data it is often inferred that the economy of protecting steel by galvanising is small if the structure is to be fully immersed in sea water. On the other hand, where for some reason a paint coating cannot be used, galvanising should be seriously considered since a more equal distribution of attack takes place on zinc-coated steel than on the uncoated metal.

Zinc is stable up to 50-55°C in soft and distilled water, but on subsequent rise in temperature the attack increases up to 65-70°C (Chap. VIII). At higher temperatures the rate of attack decreases owing to the decreasing solubility of oxygen in water and the change in nature and composition of the corrosion products. If the carbon dioxide content is greater than 27 ppm the attack is considerably reduced but below this value the rate of attack is relatively constant. Protective layers are formed more rapidly under conditions of alternate immersion.

In hot natural waters the behaviour of commercial zinc is mainly determined by the particular properties of the water. In general, the attack is greater at higher temperatures and higher oxygen concentrations. Between 70 and 80°C the attack is particularly marked and is localised in all cases. If the carbonate hardness is adequate the attack is reduced, e.g. the attack is reduced by moderate values of carbonate hardness and at 70°C both zinc and zinc with a small lithium content are comparatively stable with a dissolution rate of a 3 mdd. The rate of attack increases with decreasing hardness in the following series: normal supply water, soft water, and soft brackish water (as in coastal regions). The formation of a protective film

produces a decrease in attack with time in fresh waters. In brackish
water this decrease is only small.

In certain waters, particularly soft waters with an appreciable bicarbonate
content, troubles can arise in hot water owing to the reversal of the potential difference between iron and zinc. In such media the potential of zinc
becomes more noble than iron at temperatures above about 60°C. If this
occurs the sacrificial protection afforded at breaks in a galvanised coating is lost and attack on the basis metal can actually be stimulated.

ZIRCONIUM

Zirconium is similar to titanium in relying for its behaviour on the
presence on its surface of an inert and tenacious oxide film. It does not
suffer any appreciable attack by water at ordinary temperatures. At elevated temperatures on extended exposure to boiling water, ductile zirconium is transformed into the brittle state due to a gradual solution in the
metal of oxygen and hydrogen from the water. The solution of at least
100 ppm of hydrogen embrittles the metal. Like palladium, zirconium very
easily dissolves large amounts of hydrogen, the maximum solubility corresponding to the zirconium hydride, ZrH_2. This solubility increases with
rise in temperature.

Zirconium is very stable to chloride solutions at ordinary temperatures
and is similar to titanium in showing an exceptional stability in sea water.

Very pure zirconium was first used in nuclear reactors but now a group
of alloys, the Zircaloys, are used for this purpose. These have improved
corrosion resistance in waters above 290°C.

The resistance to corrosion in water and steam is markedly affected by
impurities such as nitrogen, aluminium and titanium. Dissolved ferric and
curpic chloride, even at small concentrations, increase the attack.

In high temperature water zirconium and many of its alloys show a period
of decreasing corrosion followed by a transition to a more rapid linear
rate. The time to the transition and the subsequent rate of attack vary
with temperature and purity. The corrosion is extremely sensitive to dissolved nitrogen; more than 40 ppm produces a rapid transition to a high
corrosion rate. Small amounts of aluminium and titanium have also a bad
effect.

Alloys such as Zircalloy 2 and 4, containing small amounts of tin, iron and
chromium, are far superior in high temperature water. Improper heat
treatment and hot working result in the precipitation of intermetallic compounds in the microstructure and give rise to poor mechanical properties
and corrosion resistance.

NOBLE METALS

The noble metals, gold, silver, platinum etc. have inherently an extremely
high corrosion stability and do not rely on the formation of any protective

oxide or other film on their surface. Owing to their high cost and low strength their use is limited to providing thin films and liners on other constructional materials. They are in general only used where no other material has sufficient resistance to attack and are rarely employed in equipment in contact with natural and treated waters. It is well, however, to bear in mind that where absence of corrosion is vital these materials may be quite economic. A very high proportion of the initial cost will be recoverable when the apparatus becomes obsolete or damaged.

Platinum anodes are used with advantage for cathodic protection installations particularly where the small size of the anode is important as, for example in the protection of pipes and pumps.

Silver solders, alloys based on the silver-copper-zinc-cadmium system are used extensively for brazing. They sometimes also contain small amounts of nickel and tin. The brazing alloy is usually cathodic to the base metal and is in general protected.

Brazed joints on austenitic stainless steel have poor resistance to chlorides. Corrosion of the solder occurs, the joint loses its strength and rust spots appear on the steel. If 3 per cent of the cadmium in the 50 per cent silver alloy is substituted by nickel the behaviour of the joint on the stainless steel is improved. The strength of the joint is retained but corrosion of the solder and the steel is not eliminated.

The production of strong joints on nickel-free chromium steels is a more difficult problem. When joined with nickel-free brazing alloy rapid corrosion of the joint interface takes place even in tap water. The corrosion is restricted to the crevice formed at the interface between the steel and the brazing alloy. Anodic attack takes place on the active steel in the crevice, resulting in loss of joint strength and development of rust spots. Corrosion may be prevented by using a brazing alloy containing nickel, e.g. 60/28/10/2 silver-copper-tin-nickel. Attack in the crevice is eliminated by a layer rich in nickel deposited at the steel-solder interface.

Chapter VI

CORROSION AT BI-METALLIC CONTACTS

INTRODUCTION

A common way in which galvanic cells can arise in industrial plant is in the use of dissimilar metals. Different metals may be chosen because corrosion resistance is important in one part of the plant while mechanical

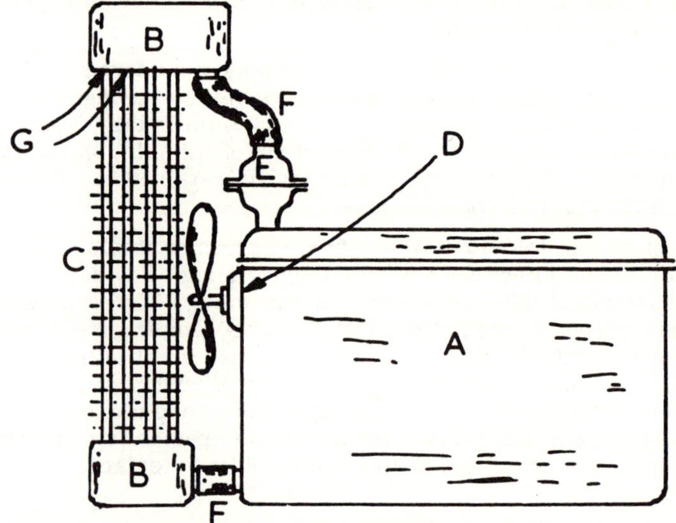

Fig. 21.   WATER COOLED ENGINE

       A - Engine block and head; cast iron or aluminium alloy. May be steel liners.
       B - Tanks; mild steel or brass.
       C - Radiator tubes; copper, tinned copper or solder-coated brass.
       D - Water pump; cast iron with hardened steel spindle or aluminium alloy with cast iron impeller.
       E - Thermostat; brass.
       F - Hoses; rubber.
       G - Soldered joints.

       N.B. In aero and tank engines, tanks and radiator tubes may be of aluminium alloy.

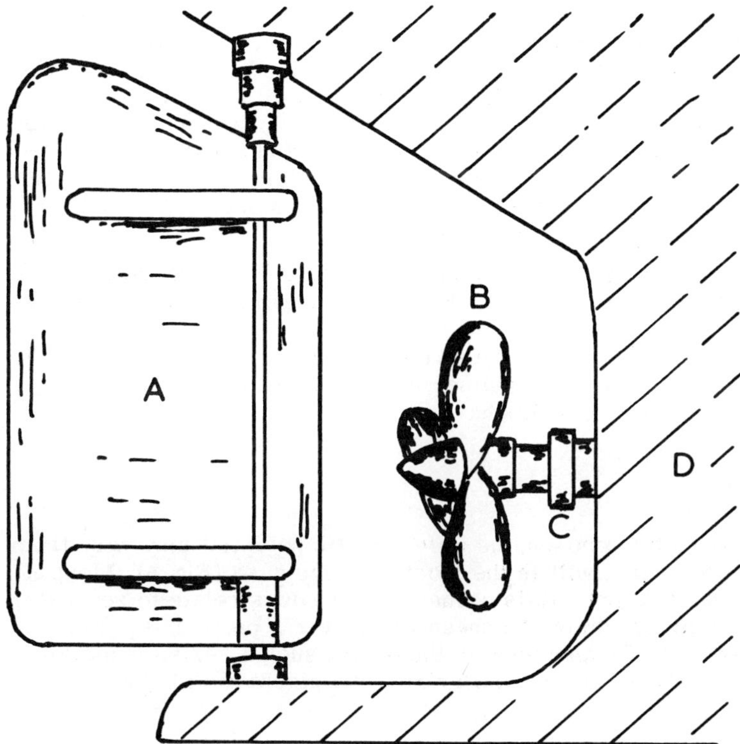

Fig. 22.   STERN OF A SHIP

    A - Rudder; mild steel with bronze fittings.
    B - Propeller; manganese brass, cast iron, bronze or
        monel.
    C - Propeller shaft; high tensile steel, bronze or
        monel.
    D - Hull; ship's plate steel.

strength is of prime concern in another. In other cases, in order to keep
the cost to a minimum, various parts of the system may be made of metals in
a readily available form.

Unfortunately even where cost can be ignored and metals are chosen
primarily for mechanical behaviour or corrosion resistance, this may be
done without due regard for the compatibility of one metal for another.
One of the best known, disastrous and expensive cases of mixed-metal
attack occurred on the yacht 'Sea Call'. No expense was spared in using
the most corrosion resistant materials available but no account was
taken of the galvanic cells formed by different combinations of metals.
The underwater part of the hull was plated with monel, a 67/28 nickel-

copper alloy, using both steel and monel rivets and the keel, stem and
rudder frame were made of steel. The designers, however, failed to take
into account the fact that the hull would act as a cathode of a huge galvanic
cell and accelerate the attack on the steel parts which formed the anode
of the cell. As a consequence of this, the yacht, which had cost half a million dollars, almost fell to pieces during its trials and it had to be scrap-
ped without its having made a single voyage. If a plain galvanised or well-
painted iron had been used instead of the expensive monel the vessel
would have been seaworthy for many years.

Two very different examples are shown in Fig. 21, the cooling system of
an internal combustion engine and in Fig. 22, the stern of a sea-going ship.

In the following pages we shall discuss the compatibility of metals and the
dangers that can arise from the indiscriminate choice and mixing of
metals with different corrosion behaviour.

GALVANIC SERIES

Depending on the metals chosen, the difference in potential may vary from
a few millivolts to over a volt in the most extreme case (Fig. 4). Most of
the data on the relative potentials of metals and alloys relate to sea water
or water with a similar chloride concentration, viz. 3 per cent sodium
chloride. Nevertheless, experience in the use of such a series leads to
the belief that metals will have similar relative potentials in other natural
waters.

These potentials range from the base value of less than −1. 3 volts for
magnesium and its alloys to noble values, up to 0. 2 volts positive to the
hydrogen electrode, given by metals such as gold and silver and by the
non-metal graphite. Stainless steel is shown to range between two positions corresponding to its normal passive state, when its potential is
close to that of copper and its alloys, and its active state, when its potential is close to that of iron and steel. Similar deviations in a negative
direction are obtained under adverse conditions with a number of other
metals such as aluminium and titanium which owe their 'elevated' position in the series to the presence of the stable oxide film on their surface.

Zinc can exhibit a more noble potential in hot waters of certain composition which can give rise to corrosion troubles on galvanised iron
(see p. 110). Copper and nickel, and their alloys with other metals, occupy
a high position in the table corresponding to their high corrosion resistance. Copper is some 0. 8 volt more noble than zinc, and this difference
is reflected to a limited extent in the relative potentials of copper-zinc
alloys, alloys with a higher copper content being more noble than alloys
with a high zinc content. Similar trends may be noted in other alloys, the
presence of the metal with the lower potential tending to lower the potential of the alloy.

Such complete information does not exist for any medium other than sea
water although much information is available on the behaviour of parti-

cular metal couples in certain natural waters. Some useful measurements on the behaviour of a number of metals and alloys in flowing hot water at 71°C in an automatic domestic water heater over a period of ten months have been reported (Fig. 23). This shows the probable direction of galvanic effects and also which materials overlap so that variations in the direction of current flow may occur. In most cases, however, the changes in potentials occurred simultaneously in the same direction and to the same amount so that the potential differences between the metals remain fairly constant. This diagram serves to emphasise the large difference in potential between galvanised iron and zinc on the one hand and copper and nickel and their alloys on the other.

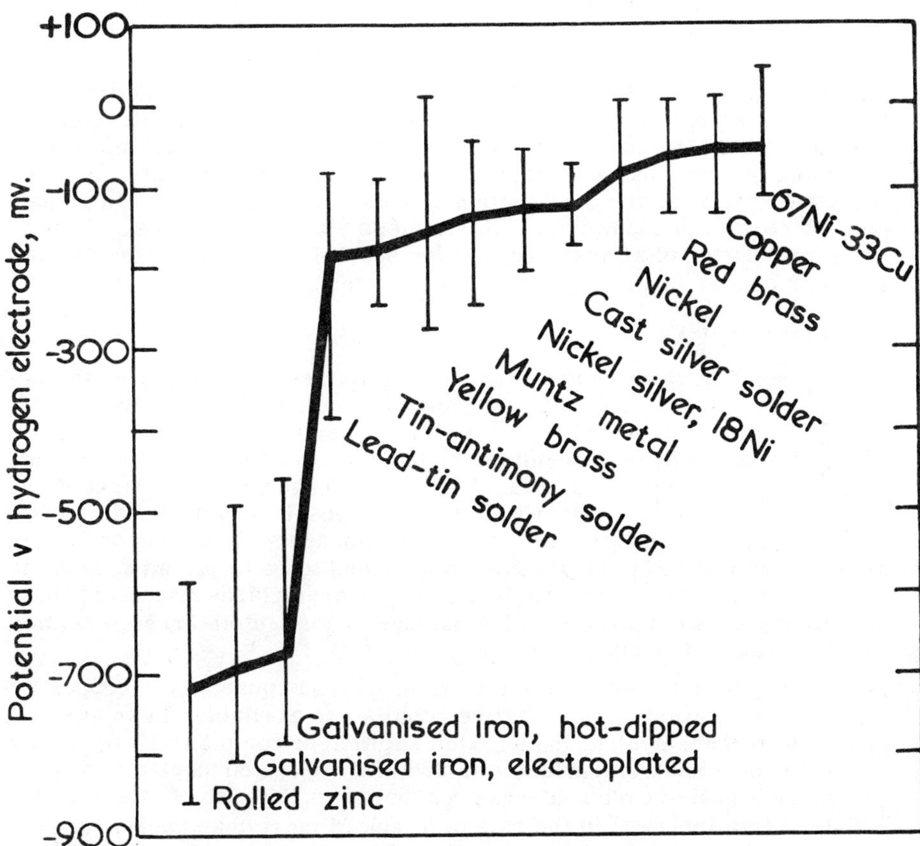

Fig. 23. GALVANIC SERIES SHOWING RANGES OF POTENTIALS OF METALS AND ALLOYS IN FLOWING HOT WATER AT 71°C (Long Island, N.Y.) Potentials measured weekly for 3 months and then monthly for a period of 10 months. (International Nickel Co.,)

On the basis of the galvanic series it might be supposed that it would be possible to predict the behaviour of any metal couple. The connection of any two metals in the series would be expected to increase the attack on the metal lower in the series and would reduce or prevent attack on the more noble metal. Several methods of assessing whether a pair of metals is compatible have been used. Use has been made of the criterion that if the metals differ in potential by more than 0.25 volts they must not be used in severe service conditions, 'quarter-volt criterion'. It is wrong to assume that the driving force in the operation of a combination of two metals is determined by the difference in their potentials on <u>open circuit</u>. When the metals are short-circuited, as when bolted or riveted together, other factors such as polarisation, relative areas and the conductivity of the medium all play an important role. According to Evans 'the quarter-volt criterion must be condemned as being completely unreliable and its pseudo-scientific flavour only makes it the more dangerous'.

The coupling of aluminium and copper, or aluminium and stainless steel, is a good example of how dangerous indiscrimate use of the galvanic series can be. The potential of stainless steel and copper are much the same and the combination of aluminium and stainless steel should give the same E.M.F. as the combination of aluminium and copper. In actual fact the contact of aluminium with stainless steel produces relatively little stimulation of attack except under special conditions whereas contact with copper invariably has disastrous results.

RELATIVE AREAS

The influence of relative areas may be easily seen by comparing the behaviour of copper plates connected together with aluminium rivets and aluminium plates fastened together with copper rivets. In the first case the total attack will be determined by the area of copper and the rate at which oxygen reaches this area. The corrosion will be concentrated on the small aluminium rivets which will be intensely attacked and will finally lead to the total destruction of these components. In the second case, although the attack on the aluminium will tend to be localised around the copper rivet the main corrosion is distributed over a large area of the aluminium plates so that the relative damage is less. Both arrangements are, however, bad and should be avoided.

Similar conditions apply to the behaviour of iron connected to copper, on which some interesting quantitative results are available. In an aerated, moderately concentrated, and agitated solution of common salt the attack on an unconnected iron surface was 350 mdd. When connected to a copper surface twenty times its own area, however, the rate of attack was 7,900 mdd, an increase in the rate of attack of more than twenty times. Again, in a 6 per cent solution of sodium chloride, the attack on a small iron surface in the middle of a large copper surface was twelve times that of a large area of iron surrounding a small copper surface.

So far we have referred only to solutions to which oxygen access is unrestricted. Sometimes the oxygen concentration is limited by the rate of transport across a relatively small liquid—air interface. In many cases

the metal will be able to utilise fully all this oxygen in its own cathodic
process. In certain circumstances it has been found that the attack on
iron in contact with copper, nickel and chromium, and also on zinc in contact with copper, are independent of the area ratio. On the other hand the
attack on zinc in contact with nickel, chromium and iron was related to
the relative areas owing to the fact that in this case appreciable evolution
of hydrogen took place.

CONDUCTIVITY OF WATER

If two metals A and B are butted together in a single plane and immersed
in water of low conductivity the attack on the anodic metal B will be confined to a narrow zone close to the junction line (Fig. 24). The total attack may be low but the corrosion will be relatively intense locally.

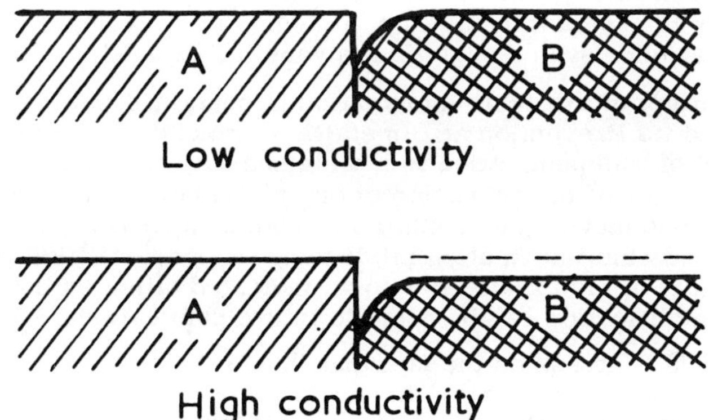

Fig. 24.   EFFECT OF CONDUCTIVITY OF WATER ON
INTENSITY OF ATTACK AT BI-METAL JUNCTION.

If, through the presence of salts in the water, the conductivity is high, the
attack is more spread out and the total corrosion will become greater.
Increase in conductivity will not in itself increase the intensity along the
marginal strip. Owing to polarisation, there is some current density
which can never be exceeded and this should be reached along the junction line whether the conductivity is high or low. Of course, if the salt reduces polarisation, or interferes with the formation of protective films,
it may increase the intensity of attack at the junction line. Thus if B is
magnesium the intensity of anodic attack will be increased by addition of
chloride to the water.

In a conducting, well-aerated medium such as sea water, the influence of
galvanic couples are shown to their greatest advantage—or disadvantage.
Engineers have for a long time appreciated that bronze propellers can
promote pitting on ships' hulls. This attack can be quite extensive since

the steel is only exposed at small breaks in the paint giving a large cathode-anode ratio and the rapid movement of the propeller will assist oxygen replenishment and thus enhance the cathodic reaction. The application of cathodic protection, e.g. by bolting zinc blocks to the stern, may counteract this effect (Chap. XI).

PRACTICAL EXPERIENCE

Clearly there are many factors which influence the decision as to whether connection of a given metal with a second metal will affect the corrosion of either. Although we go a long way on the basis of the galvanic series and the factors outlined, any deductions will be limited and uncertain since the galvanic series has been determined for sea water, and solutions of an equivalent chloride concentration, under conditions of relatively closely controlled laboratory conditions. In particular, the length of period of any such tests is necessarily limited. It seems preferable to rely on practical experience even though this is very incomplete.

A useful summary of available information is contained in a document on 'Corrosion and its Prevention at Bimetallic Contacts' (see Bibliography). The main part of this publication is a table showing at a glance whether there is any danger of the corrosion of one metal being increased by contact with a second metal. The compilers make an appeal to any designer or user who finds the information misleading in any particular application to make a report of his experiences to an indicated office so that corrections or amplification can be made to future publications.

The document lays stress on the fact that the condition of the metal may affect its behaviour, e.g. variation in corrosion potential after heat treatment of certain types of aluminium-copper alloys, influence of oxygen in maintaining the passive film on certain metals, and the influence on the potential produced by variations in alloy composition.

A much simpler scheme is the one put forward by American workers interested in materials of construction for naval aircraft. Metals and alloys are classified in four groups:

1. Magnesium and its alloys

2. Cadmium, zinc, aluminium and their alloys

3. Iron, lead, tin and their alloys (other than stainless steel)

4. Copper, chromium, nickel, silver, gold, platinum, titanium, cobalt, rhodium and their alloys. Stainless steels. Graphite

Combinations are only regarded as incompatible if the two metals come from different groups. A junction of two materials from the same group may be regarded in many respects in the same way as a combination of two parts composed of the same material. Although this appears to be less stringent than British requirements the treatment of surfaces composed of a single material is carefully prescribed and is probably suf-

ficiently elaborate to take care of cases where two materials from any
one group are present.

In the remaining sections of this chapter practical experiences of the
behaviour of metal couples are considered in order starting with the
most anodic, i.e., base metals, and progressing upwards through the
galvanic series. Although the information available is far from complete
practical experience gained is generally in those fields where the problems are more serious, hence some of the greatest dangers in the use of
bimetal couples are highlighted.

MAGNESIUM AND ITS ALLOYS

Magnesium is the most reactive of the constructional metals and is anodic
to other metals. It has little tendency to polarise in most environments
and hence will undergo severe galvanic attack if precautions are not taken.
Use is frequently made of this characteristic to protect other metals in
aggressive conducting media (Chap XI). Heavy metal impurities in magnesium can cause galvanic corrosion of the magnesium itself.

Pure aluminium is unique in being almost completely compatible with
magnesium alloys but this is not so with commercial aluminium alloys
which usually contain iron and copper. The compatibility can be restored
by addition of magnesium to the aluminium alloy. Thus rivets or screws
of 5 per cent magnesium alloy have been used extensively with magnesium
alloys for many years without causing any significant galvanic attack. In
contrast, fastenings of duralumin alloys, which contain copper, produce
corrosion.

The anodic behaviour of magnesium to aluminium alloys can be used to protect the latter but such a combination should be used with caution since the
alkali produced at the aluminium can produce corrosion of the latter
(Chap. I). Such behaviour has been reported in sea water and other saline
liquids. This galvanic attack on magnesium alloys coupled to noncom-
patible aluminium alloys in saline solutions can be inhibited by chromates,
vanadates or sulphides or mixtures of these chemicals but problems
exist in providing a continuous supply of the inhibitor to the contacting
surfaces.

Severe corrosion of magnesium alloys is caused on connecting with iron,
nickel and copper or alloys based on these metals and titanium-based
alloys are only slightly better. Screws, studs, nuts, bolts and washers
made of steel, copper or brass should preferably be plated with cadmium
or tin but zinc or chromium may also be used. Cadmium is probably the
best owing to its combination of polarisation and corrosion properties
but in sea water magnesium has been attacked at its contact with cadmium-plated steel. The use of tin electroplate is increasing because of
its superior behaviour with regard to polarisation and corrosion.

Owing to the big potential difference the most severe attack on magnesium can clearly take place when in contact with carbon. This occurs
if magnesium tubes carry a discontinuous coating of graphite derived
from the lubricant used in fabrication.

ZINC AND ITS ALLOYS

Zinc is anodic to iron and almost all the common constructional metals and protects iron, nickel, lead, tin and copper. It is in turn cathodically protected by magnesium.

In neutral or acid solution zinc is anodic to aluminium but little increase in attack on zinc is observed. On the other hand at high pH values aluminium is anodic as might be predicted from an examination of the dependence of the corrosion of the separate metals on hydrogen ion concentration (Chap I).

Zinc gives good galvanic protection to steel in sea water and fresh waters. It is most effective where the electrical resistivity is fairly low as in sea water where it is used on ships' hulls for cathodic protection.

At temperatures above about 70°C a potential reversal can occur in some waters and zinc, becoming cathodic to steel, no longer protects the latter and may even be capable of stimulating attack. The potential of zinc is influenced by water composition and if the bicarbonate content is increased to about 130 ppm it becomes more cathodic, particularly at higher temperatures. Sulphate and chloride tend to maintain zinc at a more negative potential and anodic to steel. In solutions free from oxygen zinc is anodic to steel at all temperatures. A potential reversal may also be produced by old corrosion layers on a zinc coating.

Coupling to stainless steel doubles the weight loss on zinc.

Zinc and copper or copper alloys are not in general compatible. In fresh water plumbing systems, however, brass connected to galvanised steel hot water tanks does not often cause accelerated corrosion of the tanks because the relative area of the brass is small compared to that of the tank. Where the relative areas are unfavourable, as in the use of galvanised iron couplings in a brass pipeline, troubles can occur and such practice should be avoided.

If water containing free carbon dioxide flows through a copper pipe and then through a galvanised tank serious corrosion conditions can be set up. Such water can take up traces of copper as copper bicarbonate and later deposit metallic copper on the tank or cistern. These copper-zinc micro-cells cause rapid attack on the zinc coating and this may ultimately result in perforation. A case is on record in which every galvanised tank on a housing estate failed within four years when the supply water contained 4.1 ppm free carbon dioxide and took up 0.32 ppm copper. On another estate, where the water contained only 1.1 ppm free carbon dioxide, the copper taken up only amounted to 0.03 ppm and no failures occurred.

Such micro-cells may be more dangerous than the macro-cells set up when, for example, a copper pipe is in electrical contact with a tank or cistern. It has even been claimed that deliberate connection between copper and galvanised iron can improve the situation since the zinc affords cathodic protection to the copper so that micro-cells do not develop. This would appear to be a very dubious practice, however, since in many waters the copper would only be protected over a very limited length.

In hard waters little trouble is likely since the whole installation soon becomes covered with a calareous deposit, nor is any trouble likely in waters of higher pH value where the free carbon dioxide content is low. Thus remedies for this type of attack are either the increase of the hardness and pH value by addition of lime or the aeration of the water to remove free carbon dioxide.

ALUMINIUM AND ITS ALLOYS

Aluminium is anodic to many common metals. The alloys of aluminium vary greatly among themselves although it is common in many circles to call them all 'aluminium'. Galvanic attack can take place between different aluminium-copper alloys if their copper contents differ by more than about 2. 5 per cent.

Aluminium will not be adversely affected by contact with zinc and cadmium or parts plated with these metals providing the coatings are continuous and have an adequate life.

Contact with copper and copper alloys produces rapid attack on aluminium. If such a connection is necessary, protection of the contact area is essential. Anodic oxidation of the aluminium is not sufficient to protect it from galvanic attack. On the other hand duralumin screws after anodic oxidation are suitable for connecting aluminium or alclad alloys.

Aluminium rain-water piping in electrical contact with copper has suffered rapid perforation. Insulation is not a sufficient remedy if the rain flows over the copper and then carries traces of copper salts on to the aluminium. Traces of copper taken up by waters from copper pipes also causes pitting of aluminium cooking vessels. This is particularly dangerous in vessels which cannot be adequately cleaned such as kettles containing an electrical immersion heater. It has been found experimentally that a copper concentration of only 1 part in 50 million is sufficient to cause attack if calcium bicarbonate, chloride and oxygen are also present.

With aluminium alloys containing copper, trouble at junctions is generally reduced if the material has been clad with pure aluminium which is anodic to the parent alloy. Sometimes benefit can be obtained by spraying the whole assembly with pure aluminium.

In fresh waters aluminium and its alloys can be anodic or cathodic to iron but in most cases the potential difference between the two is small and there is little influence on the attack on either. An aluminium-magnesium alloy will generally be anodic towards steel whereas an aluminium copper alloy may sometimes be cathodic. This depends on the composition and heat treatment. Experiments have been described where an alloy was distinctly cathodic to steel when 'naturally aged' and anodic when 'artificially aged'. In many natural waters stainless steels or chromium-plated steels, even with their high cathodic potentials, usually cause negligible attack on aluminium because the galvanic current is limited by polarisation.

In sea water aluminium is anodic to iron and its corrosion is increased. This effect is even more marked when connections are made to stainless steels and aluminium alloys may be destroyed.

Where connection with steel is unavoidable and it is essential to minimise the danger of corrosion, the aluminium should be anodised or the steel may be coated with either cadmium or aluminium.

Troubles arising from the contact of aluminium with steel are not necessarily avoided when the two metals are insulated one from another. If the corrosion products from the steel can reach the aluminium, accelerated attack on the latter can arise from a number of causes: alkali produced at the steel cathode can increase the pH, iron salts formed anodically may deposit iron or magnetite forming a corrosion couple and ferrous hydroxide reaching the aluminium surface will remove oxygen and set up differential aeration cells. Clearly the designer should attempt to avoid such a disposition of materials that permits corrosion products from iron to sweep or settle on light metals and alloys.

The coupling of aluminium with nickel is not good practice in most acid or salt solutions. The use of aluminium paddles with a nickel milk-pasteuriser resulted in complete protection of the nickel and rapid corrosion of the paddles. The area ratio of nickel: aluminium of 5:1 was distinctly unfavourable although when the two were not coupled the aluminium was hardly attacked.

Aluminium-lead couples should also be avoided. Paints containing lead pigments will cause severe galvanic corrosion on aluminium subsequently exposed to solutions containing small amounts of chloride or sulphate.

IRON AND ITS ALLOYS

All types of low alloy and straight carbon steels can be used together without danger of increased attack. Such steels, however, are anodic to the stainless alloys containing large amounts of chromium and nickel.

In quiescent or slowly moving sea water, the increase in corrosion of mild steel when in contact with copper, nickel, stainless steel or titanium is roughly the same but is proportional to the area of the more noble metal. On the other hand in rapid movement the increase in corrosion depends on the metal contacted, e.g. with copper or nickel the increase is appreciably larger than with stainless steel or titanium.

In considering protective metallic coatings the relative position in the galvanic series with regard to iron is vitally important. When the coating is anodic, e.g. zinc or cadmium, the protection is good so long as the area exposed is not extensive, but when the underlying iron is anodic e.g. to nickel or chromium, exposure at pores, cracks or areas of damage may lead to pitting.

The large difference of area makes steel rivets particularly susceptible to attack if used in a more cathodic metal and when used in a 14 per cent chromium steel, mild steel rivets were almost completely destroyed. In mild steel plate there is no accelerated attack due to galvanic effects but rivets are often heavily corroded because they are sites of poor paint adhesion. This corrosion can be prevented by alloying with small additions of nickel and copper thus making the rivets more noble than the surrounding mild steel. Although these additions have no effect on the inherent corrosion resistance (Chap. IV) they increase the potential sufficiently for them to enjoy some measure of cathodic protection from the surrounding plate. A fifteen-year test on alloy steel rivets in mild-steel ships' plate indicated that compared with mild steel, rivets of 2 per cent nickel steel were twice as resistant and rivets of 2 per cent nickel 1 per cent copper steel three times as resistant. No localised attack took place in the vicinity of the rivet.

In a similar manner a 2 per cent nickel-steel weld suffered less corrosion than a carbon-steel weld and did not accelerate corrosion or pitting of the mild-steel plate.

The behaviour of stainless steel depends on whether it is in the active or passive state. When in the active state stainless steels behave galvanically like iron. It is not always possible in practice to say with certainty whether stainless steeel is in the active or passive state since the presence of air, oxygen or oxidising material and the concentration, temperature, and composition of electrolyte all have an influence.

When the degree of aeration is relatively small, as in stagnant water, passive stainless steel is likely to become active particularly in a conductive water such as sea water. In such cases contact with less noble metals will have little effect on either metal; the corrosion of ferrous alloys will be increased a little and the increase will be more noticeable on more electronegative alloys such as those of aluminium. For example, in constructions of low alloy steels the use of rivets or welding rod of stainless chromium-nickel steel is possible. In such a combination the areas of stainless steel are relatively small and remain completely stable and the overall rate of corrosion of the low-alloy steel connected with it is only increased to a very small degree. On the other hand if the surface of the stainless steel is very large in comparison with the surface of low-alloy steel, cast iron, zinc or aluminium, then, although the stainless steel is not such an active cathode as copper, there is a danger of an appreciable increase in the corrosion of the more electronegative metal.

Contact with metals such as aluminium alloys, carbon steels and zinc can appreciably affect the behaviour of stainless steel in sea water. Owing to cathodic polarisation the passive state of the stainless steel cannot be maintained but at the same time the stainless steel receives some electrochemical protection. This may result in an increase in the overall weight loss but the tendency to pitting is markedly reduced. For the same reason pump impellers of stainless steel corrode less when the body is made of cast iron, although the corrosion of the latter increases a little.

When it is in the fully passive state, e.g. when in well-aerated sea water with appreciable flow, stainless steel will behave as a very positive contact material (cathode) and will be able to increase the corrosion not only of such metals as aluminium and carbon-steel but even copper and copper alloys.*

In fact, stainless steel (18-8) as cathode can cause almost as much attack on cast iron as when the latter is connected to platinum.

The combination of iron with copper does not always give rise to troubles particularly under weakly corrosive conditions, e.g. in soft mains water without carbon dioxide. If the conditions are more aggressive, e.g. in sea water, the iron is attacked more strongly since copper acts as the cathode for oxygen depolarisation as well as for other reduction reactions such as reduction of ferric or cupric ions. Troubles occur frequently in hot water installations in which copper coils are situated in iron boilers or iron or galvanised tubes are connected to copper. Troubles by direct contact action are usually limited to the vicinity of the connection. In flowing water containing oxygen the dissolution of iron remains the same even if three quarters of its surface is covered with copper although the intensity of attack will be four times as great.

On the whole, combination of stainless steel and copper is unfavourable and damage to both metals can occur. Thus, in sea water, contact of a large area of copper or copper alloy with a relatively small area of stainless steel is dangerous for the latter. In the presence of appreciable amounts of chloride the stainless steel becomes activated and is then anodic to the copper and its corrosion is thereby increased. On the other hand, contact of stainless steel with small components of copper or copper alloys is dangerous for the copper alloys. In this case the stainless steel probably attains a stable cathodic state in relation to the copper.

Among the examples of dangerous contact corrosion the use of steel rivets in a monel hull has already been mentioned. Nickel accelerates attack on mild steel and cast iron, the extent depending on the acidity of the solution, the degree of aeration and the relative ratios of the areas. The difference in potential in distilled water is 0.22 volts while in a hard supply water it may be 0.31 volts.

The polarity between nickel and stainless steel depends on whether the stainless steel is in the active or passive state but in many neutral and alkaline solutions the two materials may be used together.

In salt solutions and inorganic acids iron is less noble than tin but in fruit acids the reverse is the case, which enables tin plate to be used in the canning industry. The slightest change in conditions can lead to a change in polarity between tin and iron. The change centres around the oxygen availability and the metal ion concentration that builds up at the metal surface.

Lead is close to iron in electrochemical character but is usually cathodic.

---

* The passive state of stainless steel refers to the electrode potential established on the metal under conditions of rapid movement of well-aerated sea water. On the other hand, the active state refers to the potential established in weakly-aerated stagnant sea water.

When cast iron undergoes graphitic corrosion it becomes more noble than
the unattacked metal. Austenitic cast irons are not appreciably affected
by contact with cast iron coated with graphite. For this reason it is
advisable to use austenitic cast iron for replacement impellers in centrifugal pumps with iron casings since new grey iron impellers could be
rapidly attacked by galvanic action.

Graphitised cast iron may also become more noble than bronze alloys.
This effect could cause difficulties in the use of cast iron valves with
bronze seats and stems.

COPPER AND ITS ALLOYS

In most applications copper alloys can be safely used in contact with one
another. Copper alloys containing little or no zinc are less inclined to be
attacked when coupled to copper. On the other hand large copper surfaces
can increase the corrosion of brass in sea water. Condenser tubes of
high-copper alloys are commonly used in contact with tube plates of high-
zinc brass in heat exchangers handling either fresh or salt water. There
is little danger here since the tube plate, of thicker section, is the more
negative alloy (Fig. 4).

In strong chloride solutions adherent corrosion products on copper can
ennoble its potential by about 200 millivolts. This is important in preheaters where local deposits may increase the corrosion of the copper by
50 to 80 per cent.

Copper and copper alloys are cathodic to most common metals and can
stimulate attack to an extent determined by the areas of surfaces. Copper can be safely connected to lead, nickel, or tin exposed to water and
salt solutions and to iron when the area of the latter metal is large compared with the area of copper. As neutral solutions become more concentrated in salts, and therefore better electrolytes, the danger in coupling copper to other metals becomes more pronounced. When the area of
copper is relatively large it is best insulated from contact with iron,
steel, tin and lead and preferably not used with aluminium or zinc. In
most water systems copper and lead can be connected together since the
lead becomes covered with a protective layer, e.g. lead sulphate.

The coupling of nickel-copper alloys, copper, brass or bronze with nickel
is usually safe in practice since only small potential differences of 0.05
volts or less are developed in neutral solutions. In more acid solutions
such combinations are best avoided since the nickel will usually be anodic
and its corrosion rate may be increased.

Copper-nickel alloys are usually cathodic to the common metals but may
be connected to other copper alloys and under many conditions may even
be connected to steel. Caution is required before connecting to aluminium
and lead. Monel coupled to soft solder in hot tap water at 93°C caused
accelerated corrosion of the solder when the ratio of the areas was
150:1. Similar tests in water at 65°C failed to give any accelerated attack. Silver solder is more compatible than soft solder.

CARBON

Graphite stimulates corrosion of most metals and markedly increases the dissolution of metals such as iron, zinc and aluminium. The corrosion of nickel is unaffected and both stainless steel and titanium are favourably influenced since they are polarised anodically. In weakly acid solutions, however, the attack on stainless steel may be increased.

Graphite in packing glands has caused trouble and where possible mica is to be preferred.

PREVENTION OF GALVANIC ATTACK

Where conditions are such that dissimilar metals cannot be avoided certain precautions can be taken during design and assembly to diminish galvanic effects. The most important point to remember is that all the metals should be electrically insulated from each other by the use of plastic inserts or sleeves. Even in the absence of insulation the use of a plastic washer can be an advantage since the length of the electrolyte path is increased.

It is advisable to fill crevices between dissimilar plates with a jointing compound and this may also serve to prevent crevice corrosion. Such compounds are more effective when used as an impregnation on a rot-proof fabric placed between the mating surfaces.

Outer surfaces around joints should receive very careful painting and, where one of the metals is non-ferrous, a chromate paint will usually be more suitable than one of the red-lead type. In painting, it is more important that both metals or the cathodic metal should be painted since painting of the anodic metal alone may lead to intense attack at adventitious breaks in the coating.

In using a metallic coating over the whole assembly the metal should be less noble than either metal, e.g. aluminium or zinc, but bad effects can sometimes be avoided by coating the first metal with the second.

In heat exchange equipment using copper coils with, say, a galvanised tank, dangers from copper dissolution may be avoided by nickel-plating the coils. These can then with advantage be insulated electrically from the tank.

As already noted chemical or anodic treatment is not generally adequate.

Another method advocated is the use of intermediate metals. In this, the object is to reduce the variation in potential between any pair of metals in contact. Thus where sulphate is present in the water, lead may be inserted between metals higher and lower in the galvanic series, the lead becoming coated with a protective layer of lead sulphate. Clearly this method will only work satisfactorily in waters of only moderate conductivity where the throwing power is low.

Inhibitors may also be used, particularly those which increase the polarisation of the cathode and make the reduction of oxygen at this electrode more difficult. Where inhibitors are used in mixed-metal systems it is found that a much higher concentration is required for effective protection than would be the case if only a single metal were used. This indicates the desirability of avoiding mixed metals in such systems.

Chapter VII

METALLURGICAL AND MECHANICAL FACTORS

The conditions necessary for the onset of corrosion are often provided by heterogeneities which may already exist on or within the metal or alloy or which may be created by external factors. Purity, of itself, is no guarantee of corrosion resistance. Indeed, while aluminium increases in resistance with increase in purity, iron is no more resistant in the pure state than is cast iron or steel. In the first case the resistance is provided by the protective oxide formed on the metal while in the latter case the lack of resistance is due to the ease of initiation of the reactions necessary for corrosion.

Alloys vary widely in composition, complexity and heterogeneity but even in simple binary alloys, if the constituent metals are widely separated in the galvanic series, selective attack may occur. The less noble metal is removed leaving behind the more noble metal, often combined with corrosion product, thus producing a material of poor mechanical properties. In other cases the precipitation of minor constituents or intermetallic compounds may create grain boundaries more anodic or cathodic than the surrounding grains or may cause a depletion of the more noble constituent in the grain boundary region rendering the latter more susceptible to attack. This intergranular attack occurs in a wide variety of alloys. Apart from the purely structural features, potential corrosion conditions are created by mechanical and constructional factors. Processes during manufacture such as bending, drawing, welding and riveting all inevitably introduce stresses which put the metal in a less resistant condition. Anodic and cathodic areas are caused by differing stress levels or the local breakdown of films and oxides. Furthermore, during such welding or bending operations the application of heat may result in modification of the metal structure.

A most dangerous situation can arise when mechanical stress is associated with corrosive attack, which is very localised. Rapid failure may be caused by stress corrosion cracking and corrosion fatigue depending on whether the stress is static or fluctuating. The essential conditions of operation of plant may create the very stresses it is desirable to avoid.

The conjoint action of corrosive influences and mechanical damage when metal is subject to the impact of certain constituents in solution, is known as corrosion erosion. These constituents may be solid particles such as sand or bubbles of air, causing impingement attack, or cavities, which on collapse produce cavitation. Bad design and manufacture are sometimes responsible for the creation of conditions which may well increase the danger from any of the factors mentioned.

SELECTIVE ATTACK

GRAPHITISATION

Cast iron has the most complex structure of any alloy used in contact with
water. It consists of a mixture of ferrite, Fe, cementite or iron carbide,
Fe_3C, graphite, C, iron phosphide, Fe_3P, iron sulphide, FeS, and manganese
sulphide, MnS, all of which have different galvanic potentials and hence
provide many possible corrosion cells. Initially the corrosion of cast iron
is similar to that of mild steel and in many waters continues in a like
manner. In certain waters, however, graphitisation occurs, some of the
products of attack remaining where they are formed so that the affected
part consists essentially of graphite flakes and cementite particles
embedded in corrosion product. This results in the formation of a soft
porous mass on the surface of the original iron.

When such a porous mass is dried a spontaneous reaction may occur, presumably involving the very rapid oxidation of the very finely divided
material in the presence of water and carbon. This reaction can be intense
enough to raise the temperature to red heat. This happened when some
cast-iron cannon balls were recovered from a sunken warship where they
had lain exposed to sea water for over 100 years.

Although a water pipe which is severely graphitised (Pl. 13), is very weak
and easily fractured it retains its original shape and will still hold water.
On the other hand if the attack is very rapid and localised quite small
pressures are sufficient to blow out the soft plug-like mass.

No accurate data exist on the rate of attack although, in laboratory tests in
distilled water, the attack is comparable to that of steel under the same
conditions. The attack is favoured by low pH values and high conductivity.
Gases such as hydrogen sulphide are particularly undesirable since they
increase the acidity.

DEZINCIFICATION

In the selective corrosion of brasses the original alloy is converted into a
spongy mass of copper which has poor mechanical strength. This type of
attack is most rapid in acid and strongly alkaline water and is accelerated
by poor aeration and high temperature. The most dangerous natural
waters are soft waters containing free carbon dioxide, brackish waters
and sea water. There is evidence that iron and manganese in the brass
accelerate this form of attack.

Two theories on the mechanism exist; either that the zinc is dissolved
preferentially leaving the copper behind, or, the more favoured, that both
metals dissolve and the copper is subsequently redeposited. The attack
is encouraged by lack of water movement which helps to keep the copper
ions on the metal surface.

The zinc-rich brasses, i.e. the $\alpha\beta$ and β brasses, are the most susceptible
to this form of attack. The copper-rich α phase is usually cathodic to the
zinc-rich β phase. In the dezincification of duplex brasses the anodic β

phase dissolves first and then the α phase but both processes can operate simultaneously as each is accelerated by the copper produced. A continuous network of the β phase is known to be harmful particularly if it comprises less than 10 per cent of the microstructure.

α-brasses with a zinc content of 15 per cent or less are very resistant to this form of attack. The susceptibility increases with the zinc content and both brass with 23 per cent zinc and aluminium brass containing 22 per cent zinc and 2 per cent aluminium are often affected in soft waters. In the α brasses the attack usually starts at points rich in zinc, e.g. specks of β-phase at grain boundaries.

There is no reliable method of completely inhibiting the two-phase alloys but tin has some inhibiting action particularly in casting alloys. Thus naval brass and manganese brass with about 1 per cent tin have a reasonably good resistance and are attacked more slowly than the straight brass in sea water but in most fresh waters the difference is very slight. Complex two-phase alloys of high strength and containing tin, aluminium, iron and manganese have a reasonable resistance but are not immune.

Small additions of arsenic (0.02 to 0.06 per cent), antimony or phosphorus are largely successful in preventing dezincification of α-brasses in fresh waters and sea water but do not prevent dezincification of the β phase and are thus largely ineffective in duplex brasses. Of the three additives, arsenic is the most widely used and is the only inhibitor added to aluminium brass. If it exceeds 0.1 per cent, or if the phosphorus content is greater than 0.02 per cent, aluminium brass is susceptible to intergranular attack.

Owing to their good resistance to dezincification in soft waters, red brass (up to 15 per cent zinc), Admiralty brass (29 zinc and 1 tin), arsenical aluminium brass (0.05 per cent arsenic) and arsenical muntz metal (0.025 per cent arsenic) are widely used. If conditions are more severe, as in waters with a high carbon dioxide content, it may be necessary to apply a protective tin coating. In the more aggressive conditions in sea water brass may be replaced by aluminium bronze but even then the possible occurrence of dealuminification should be considered. The use of cathodic protection by zinc protector blocks not only prevents dezincification but also arrests the action beneath the affected layer.

DEALUMINIFICATION

In waters of high alkalinity, sea water, or in alkaline environments aluminium bronzes are known to undergo a form of selective attack known as dealuminification. These cases of severe preferential attack are associated with a continuous network of the γ-phase in the microstructure. Such a dangerous network may be avoided by quenching the alloy from 600°C, but this is not always practicable and will, in any case be nullified by any subsequent heating such as will occur in welding, brazing or operation at elevated temperatures. An alternative and preferred method is to avoid a continuous network by restricting the aluminium content of such alloys to the range 8.4 to 9.1 per cent and when possible to combine this with suitable heat treatment.

A third alternative is to use an aluminium-silicon bronze e.g. 6.5 per cent aluminium, 2 per cent silicon and 1 per cent iron, which has good corrosion resistance irrespective of the method of heat treatment and is easily welded.

STRESS CORROSION

The combined effect of corrosion and static stress is able to produce a form of attack in which cracks develop at right angles to the direction of the applied stress. The essential conditions for stress corrosion are the presence of a corrosive medium and stresses, which must be of a tensile nature, on the surface of the metal. The stress needed is considerable and residual or concealed stresses within the metal are of far greater importance than externally applied stress which is easily controlled. Such internal stresses are produced by differential cooling or deformation and arise during cold-working, deformation, heat-treatment or misalignment during assembly.

Although the environment is important and often specific to a metal or alloy, e.g. ammonia and nitrogen compounds in the case of brass, in the absence of stress the attack would often only be superficial. The time taken for such attack to become appreciable may vary from several minutes to several years.

The border-lines between corrosion fatigue, stress corrosion cracking and the effects of stress on general corrosion are not sharp, e.g. stress corrosion cracks may be widened by subsequent crevice corrosion. Stress corrosion and corrosion fatigue are similar in the impairment of the mechanical properties produced and the fact that the cracks are transverse to the stress. However, in stress corrosion the stress is usually internal and the crack is intercrystalline, following a path similar to that of corrosion, while in corrosion fatigue failure is due to any stress and the crack is often transcrystalline and following a path similar to mechanical fatigue.

Stress corrosion does not occur with pure metals—pure aluminium is quite resistant—nor with certain homogeneous alloys. The materials susceptible to stress corrosion may be divided into two main classes. In the first, attack occurs following the formation of intercrystalline precipitates and only intercrystalline cracks appear, e.g. aluminium alloys containing zinc and magnesium. In the second class are homogeneous alloys which are liable to stress corrosion cracking and in which both inter- and transcrystalline attack may take place but rarely together, e.g. season cracking of brass and magnesium alloys. In the first class the tendency to cracking may be eliminated by suitable heat treatment and change in composition but with the homogeneous alloys there is no method of reducing sensitivity to cracking.

In waters which contain chloride or acid, aluminium alloys may be affected by stress corrosion. Aluminium-magnesium alloys which contain more than about 5 per cent magnesium become sensitized to stress corrosion cracking by the formation of precipitates at grain boundaries. This is due to the fact that in high-magnesium alloys the solid is in a metastable state and a relatively slight rise in temperature of 50 to 60°C leads to the decomposition of the solid solution and separation of the β phase, Mg_2Al_3, which is anodic. Grooves and pits form due to preferential etching of the precipitated β-phase or of magnesium segregated at the grain boundary. The grain boundary is anodic, of small area compared with the cathode and provides an easily corrodible path. The subsequent intergranular attack may be severe even under mildly corrosive conditions and lack of continuity in the precipitate will not necessarily eliminate cracking.

Increasing cold work produces more rapid precipitation at the grain boundaries hence the stress corrosion life is decreased. After large amounts of cold work, ageing produces a tendency to exfoliation or layer corrosion irrespective of whether the alloy is stressed or not.

Pure magnesium, magnesium-manganese alloy and magnesium casting alloys are resistant to cracking up to or beyond their tensile yield strength. In the wrought alloys the sensitivity to cracking, which is predominantly transcrystalline, increases with the aluminium and zinc content and additions of lead, tin and cadmium have little effect. In alloys with low aluminium and zinc contents, residual stresses are gradually relieved by creep. Cold-rolled, fully annealed materials are more affected than fully precipitated materials. Welded seams are prone to cracking owing to the high stress and the influence of the annealed area near the weld.

Aluminium-copper alloys with 4 per cent copper form a uniform solid solution when rapidly quenched from the temperature of solution heat treatment. The electrode potential of the grains and grain boundaries have the same value and there is no local corrosion. Stress corrosion can be caused by insufficient tempering or by heating above 100°C when precipitation occurs at the grain boundaries. The precipitate of $CuAl_2$ is roughly 0.1 V more noble and the adjacent zone, impoverished in copper, slightly less noble than the grain. According to an alternative hypothesis it is considered that the intermetallic compound is less chemically stable and preferentially dissolves owing to its electrochemical heterogeneity and intercrystalline attack proceeds by the action of corrosion cells of submicroscopic dimensions.

The ferritic stainless steels always crack in an intercrystalline manner but in the austenitic steels the cracking may be intergranular, transgranular or a mixture of the two. The intergranular mode of cracking is associated with the precipitation of chromium carbide in the grain boundaries leaving the surrounding metal deficient in chromium. This network with a lower chromium content is anodic to the rest of the metal and so conditions are created for rapid intergranular corrosion. Sensitisation to this form of grain-boundary attack can occur in welding and in the relief of stress in fabricated parts. The extent of the effect varies according to the thickness of the metal, the method and rate of welding and the heat input. The sensitivity decreases according to the welding method in the series oxyacetylene welding, arc welding to spot welding. Carbide precipitation can be eliminated by heating to a temperature in the range 980 to 1200°C, to enable the carbon to go back into solution, followed by a rapid quench. This is not easy to do and, even when it is possible, care has to be taken to avoid quenching stresses.

An alternative method is to control the carbon content at a sufficiently low value. The carbon content in austenitic stainless steels may be 0.08 per cent or more and can be lowered to 0.03 per cent or less. It is difficult to remove the carbon completely and the best practice is to reduce the carbon to as low a figure as possible and then to add elements which have a greater affinity for carbon than chromium. The two metals most commonly used for the stabilisation of steels are titanium and niobium. The titanium content should be at least 5 times and the niobium content at least 10 times the carbon content. A mixture of niobium and tantalum is sometimes used.

It is considered that nitrogen in the steel makes it more susceptible to cracking because it interferes with stabilisation by forming nitrides and must be allowed for in making the stabilising additions.

Cracking occurs in two stages, an induction period in which the reaction is largely electrochemical leading to corrosion pits and the initiation of cracks, and a later stage in which the cracks propagate.

There is no satisfactory theory and understanding of transgranular cracking although an explanation has been put forward that it is due to the presence of strain-induced martensite. The resistance to attack is rarely improved by heat treatment and it cannot be prevented by minor changes in composition. The only methods of controlling it are by stress relief annealing and limitation of applied stresses. In the absence of applied stress, only intercrystalline corrosion is active.

Season cracking is a particular type of stress corrosion of brass usually associated with atmospheric exposure but which may occur in immersed conditions. It was originally observed in the disintegration of cartridge cases produced from deep-drawing of single-phase brass. The stress required to produce cracking is rarely less than 12, 000 psi and cracking takes place anywhere between 12, 000 to 20, 000 psi. It is usually associated with the presence of ammonia and ammonium salts, amines and cyanides. An interesting case in which this form of attack occurred in a hard-rolled brass shutter in chlorinated water is shown in Plate 14.

Stress corrosion of brass proceeds in both an intergranular and transgranular manner. If the cracks are more of a transgranular nature this indicates that mechanical factors have a relatively greater influence. This happens if brasses have been deformed or have been subjected to large tensile loads and occurs in media of comparatively low corrosivity, e.g. under natural exposure conditions. Brass which has been annealed or is more moderately stressed develops mainly intergranular cracks.

The surface tension of copper is greater than that of zinc and thus zinc will tend to be positively adsorbed at grain boundaries. This increased local concentration creates a continuous surface of zinc atoms, e.g. at grain boundaries or slip planes, enabling the attack to penetrate into the metal and cause corrosion cracks. Suitable conditions are created if the medium has a selective action on the zinc or on a less noble phase with a higher zinc content. The presence of tensile stresses, which gradually widen the cracks, permits the attack to proceed along the surfaces rich in zinc.

The tendency to season cracking increases with increase in the concentration of zinc and special brasses behave like straight zinc brasses. The α brasses, particularly alloys with a zinc content of 20 per cent or less have the lowest tendency to cracking. They are made more resistant by the addition of about 0.5 per cent silicon, although this reduces the ductility somewhat, or nickel, tin and phosphorus.

ALLEVIATION OF STRESS CORROSION

The remedies used for the alleviation of stress corrosion cracking depend on modification of the metal, protecting it against corrosion, making the medium less corrosive or a combination of all three.

It is desirable to relieve internal stress by closely controlled heat-treatment, e.g. homogenisation of aluminium-magnesium alloys at temperatures between 200 and 300°C. This leads to equilibrium distribution of the anodic phase over the whole grain and the tendency to intercrystalline corrosion is thereby eliminated. If tensile stresses cannot be removed, much can be done to improve matters by producing compressive stresses in the surface by shot-peening and in some cases by rolling.

Dangerous precipitations may be prevented by additions to the alloy, e.g. chromium to aluminium alloys and titanium and niobium in the stabilisation of stainless steels.

The structure of the alloy can be modified sufficiently by slightly altering the composition as is done by lowering the zinc content of brass or the magnesium content of aluminium-magnesium alloy.

It is often feasible to protect the metal by the application of coatings or paints. Any anodic coating such as zinc will prove beneficial in protecting the parent metal and, in the case of aluminium alloys, it is useful to clad with pure aluminium. Cathodic coatings which are not continuous may make conditions worse. If it is considered that such a coating must be used particular care is necessary and a thick deposit must certainly be applied. Painting may be sufficient and, indeed, the life of magnesium alloys has been increased four or five times by the use of a good painting scheme.

The corrosion can be reduced by the use of inhibitors, e.g. in caustic embrittlement, or, in extreme cases, by the use of cathodic protection. Anodic inhibitors such as chromate may well increase stress corrosion because if the concentration drops below that required for complete protection, intense attack occurs at the anodic points which become stress raisers.

CAUSTIC CRACKING

Caustic cracking is a particular form of cracking, most commonly inter-granular but occasionally transgranular, met with in boilers where steel is exposed to caustic conditions. The concentration of caustic soda necessary for this attack is a matter of argument, the figures quoted ranging from five per cent upwards.

For this type of attack to occur the boiler water must contain substances such as hydroxide which can become concentrated in contact with the steel. This takes place in joints, seams or cracks into which the boiler water leaks slowly and becomes highly concentrated. A further require-ment is that high tensile stresses must be present. The cracking usually takes place along lines of rivets in riveted drums and at tube ends and any leaks which cannot be easily stopped by caulking or re-rolling should be viewed with suspicion.

The attack may be avoided by eliminating crevices, e.g. by using welding procedures instead of riveting. Where it is present it may be controlled by addition of inhibitors to the boiler water or by the use of feed-water treatment which does not give any free hydroxide.

NITRATE CRACKING

When exposed to nitrate, particularly at high temperatures, mild steel is susceptible to cracking. Although transgranular cracks have been reported it is more usual for the cracks to follow the grain boundaries and, in susceptible steels, these are invariably well-defined. The cracking is caused by a combination of the anodic nature of the grain boundaries and the specific action of the corroding medium. A grain boundary is a region which is inherently distorted to a degree and this is increased by the pre-sence of carbide particles and when the material is cold worked or stress-ed. The cementite or carbide particles are slightly cathodic to the sur-rounding ferrite and, furthermore, in the boundary region the ferrite lat-tice is highly distorted. These factors make the ferrite an active material and it suffers anodic attack.

CORROSION FATIGUE

When subjected to a sufficient number of cyclic stresses in air at a high enough load, a metal will crack by fatigue. Under corrosive conditions the same metal will fail at stresses well below the fatigue limit in air. This conjoint action of corrosion and repeated stress is known as corrosion fatigue. For a carbon steel of 44 tons/in^2 ultimate strength, the endurance limit for the same number of stress cycles is 45 per cent in air, 25 per cent in fresh water and 15 per cent in sea water polluted with hydrogen sulphide.

The corrosion fatigue properties of a metal depend on its corrosion resistance in the particular environment. Although heat treatment and cold work may improve the mechanical properties and endurance limit in air, the resistance to corrosion fatigue is only improved when the corrosive conditions are mild. Thus although a metal has a higher fatigue resistance when cold worked it is still susceptible to corrosion fatigue. Alloying or changes in composition are of benefit only when the corrosion resistance is increased.

The rate of corrosion is often retarded by the primary oxide film on the metal. These films generally have low strength and ductility and are ruptured, or rendered more permeable, by the action of cyclic stresses. Corrosion proceeds at such breaks, often forming sharp pits which act as stress raisers of unusually high intensity. In acid solution the formation of films is less likely but the cyclic stresses may then produce distorted metal which is very susceptible to corrosion. Waters which cause localised attack, e.g. sea water, brackish water and some well waters, are more dangerous than acid condensate.

Corrosion fatigue occurs in two stages. In the first stage pitting and crack formation takes place to such an extent that failure would occur even if the corrosive environment were now removed. After cleaning, fatigue cracks are usually visible at the bottom of the corrosion pits. These pits may be formed prior to the application of alternating stresses or simultaneously, in which case pitting is more rapid. The subsequent stage is essentially an ordinary fatigue stage for a notched specimen when the crack at the base of the pit develops until fracture occurs.

The corrosion fatigue cracks are predominantly transgranular and are normal to the direction of stress. This is further evidence that corrosion fatigue and fatigue are essentially a similar phenomenon. The existence of many cracks is characteristic of corrosion fatigue and distinguishes it from fatigue in air where there is rarely more than one crack. The large number of cracks often gives the fracture a serrated appearance.

As already explained fatigue fractures are easily recognised by having two distinct zones, one smooth (produced during corrosion fatigue), and one fibrous (produced by the final tensile fracture) (Fig. 17).

This type of attack is found in many metals and alloys. With iron and steel there is a drop in fatigue limit but alloy steels are slightly better than carbon steels and the annealed state is slightly better than the heat-treated condition. Nitrided steels behave remarkably well under conditions of corrosion fatigue.

Nickel parts exposed to cyclic stresses, e.g. springs and rotating shafts, can suffer corrosion fatigue.

The fatigue strength of copper is not affected by water as is that of a number of brasses and bronzes. Aluminium bronze loses strength after previous exposure to mains water, but aluminium-nickel bronze, on the other hand, has a high fatigue resistance. Lead is unique in that in both normal and corrosion fatigue the cracks are intergranular.

PREVENTION OF CORROSION FATIGUE

The methods of preventing or reducing corrosion fatigue are similar to those used in the case of stress corrosion but, in general, they are less successful. Inhibitors in the water are beneficial but since it is sometimes difficult to ensure that the concentration is adequate at shielded points it is preferable to use cathodic inhibitors (Chap IX). Nevertheless experiments have shown that 200 ppm sodium dichromate will completely prevent the corrosion fatigue of steel in water containing 25 ppm of both sodium chloride and sodium sulphate. In 3.0 per cent sodium chloride a concentration of dichromate of 1.6 per cent was ineffective but damage was prevented by 0.25 per cent sodium chromate. If localised corrosion occurs due to shielding of the surface the inhibitors are not effective in stopping corrosion fatigue.

In high-chloride solutions better protection is afforded by zinc chromate than the equivalent concentration of potassium chromate. This is attributed to the ability of the zinc ion to act as a cathodic inhibitor. Increasing the pH value from 6.5 to 7.6 is found to be slightly beneficial at high stresses but not at low stresses.

Dangerous tensile stresses may be replaced by compressive stresses in the surface of the metal, e.g. by shot-peening.

Most of the surface coatings may be used according to circumstances. Cadmium gives marked protection to steel against corrosion fatigue in hard mains water. With a fatigue limit in air at 27 tons/in^2, the apparent limits of bare steel in water (10^7 reversals at 1750 rpm) was 9 tons/in^2 and for cadmium-plated steel, 21 tons/in^2. It is best to avoid coatings which are applied hot, such as hot-dip galvanising, since the brittle, intermediate intermetallic layer would be a disadvantage.

Owing to the difficulty of securing completely non-porous coatings, metals such as copper and nickel, which are cathodic to iron, should be avoided. In immersed conditions the best coating materials are zinc, sprayed, electrodeposited or applied as zinc-rich paint; cadmium and, in the non-metallic field, coatings of resistant paints, particularly flexible ones like chlorinated rubber, and many of the plastics.

CORROSION-EROSION AND IMPINGEMENT

In turbulent flow a combination of the mechanical damage produced by the impingement of gas, solid or liquid on a metal surface and the inherent corrodibility of the metal may give rise to corrosion-erosion and impingement attack. The impacts on the metal surface of such materials damage the protective films already present or prevent the formation of any restraining film. A common cause of impingement attack is the presence of bubbles of air large enough to break up on impact with the metal surface releasing sufficient energy to disrupt locally any films or corrosion product. It is common for this action to occur repeatedly at the same point. Below a certain size the bubbles bounce off the surface without causing any damage and are relatively harmless. The situation is actually rather more complex in that the oxygen in the air bubbles may also act beneficially in helping to keep any oxide films in repair. Thus the corrosion potential may move sharply in a negative direction, indicating mechanical destruction of the film, or in a positive direction, indicating film repair.

This form of attack can be produced also by entrained solids of an abrasive character such as sand and silt and even by the liquid itself if the velocity is sufficiently high. Impingement by air bubbles produces smooth-sided pits, free of corrosion product and often undercut on the down-stream side (Fig. 25). Impingement pits are either tear-drop in shape or, if the turbulence is produced by particles of matter on the metal surface, in the form of horse-shoes. Abrasion by solid particles, on the other hand, usually gives uniform thinning, particularly at the bottom of a tube.

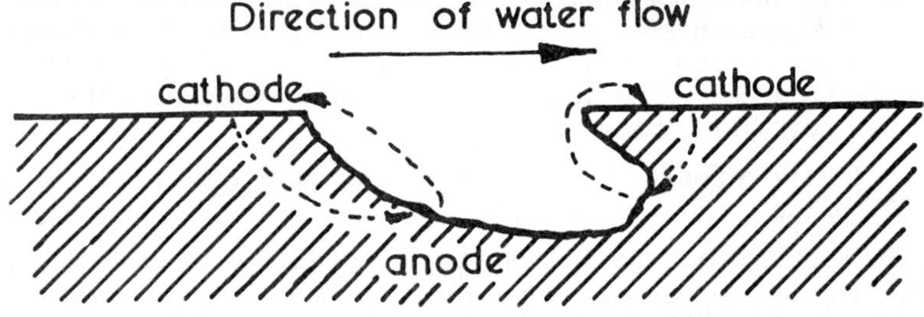

Fig. 25.   CROSS-SECTION OF AN IMPINGEMENT PIT

Copper is susceptible to this form of attack when the flow speed is high enough and the water does not form a protective scale, e.g. soft water with appreciable amounts of free carbon dioxide. Such damage can occur to ball-valve seatings. In rapidly-flowing sea water, this type of failure can be disastrous and although beryllium copper has a better resistance under such conditions it is not as good as other alloys.

Unmodified brasses do not stand up well in sea water and a number of fresh waters. Resistance to impingement increases with the zinc content and is greater for yellow than for red brass. The ability of aluminium to form protective films is conferred on alloys containing relatively small amounts of aluminium. Aluminium brass is resistant to impingement attack owing to the formation of a film more resistant to damage and better able to repair itself. This alloy is not, however, very satisfactory in polluted harbour waters and its resistance is considerably impaired when subjected to continuous scouring by sand. In polluted water containing ammonia the aluminium bronzes are susceptible to intergranular attack.

Other materials which have excellent resistance but are more expensive are the 70/30 copper-nickel alloys. It has been found that small amounts of iron improve their resistance to impingement and at the present time it is customary to add 0.4 to 1.0 per cent iron and 0.5 to 1.5 per cent manganese. An alloy with 2 per cent of both iron and manganese has excellent resistance to abrasion by silt while having a corrosion resistance comparable with a standard alloy with 0. 7 per cent iron and 1. 0 per cent manganese. The resistance in polluted water is, however, lowered. When there was a shortage of nickel a search for a material with less nickel and in which iron plays the major role resulted in the development of an alloy with only 5 per cent nickel and 1 per cent iron as a compromise between resistance to impingement and ease of fabrication. This alloy was satisfactory up to flow velocities of about 5 ft/sec. At flow velocities between 5 and 10 ft/sec this alloy was found to be satisfactory only if no additional local turbulence was created by poor geometry. Under comparable circumstances 90/10 copper-nickel and 70/30 copper-nickel containing iron and manganese undergo very little attack.

In sea water and other saline solutions, high-tin bronzes with 10 to 12. 5 per cent tin are only slightly inferior to the 30 per cent nickel alloy in their resistance to impingement attack and are superior where there is abrasion by silt. The tin bronzes are, however, less resistant to deposit attack.

CAVITATION

Cavitation can occur on any material which is exposed to water flowing with high velocity at medium or low pressures. Under highly turbulent conditions, pockets of vapour form at regions where the pressure is low and when subsequently the water passes quickly into a higher pressure section these cavities collapse suddenly. The shock pressures may reach several hundred atmospheres over a limited area of the metal surface which is sufficient to cause mechanical damage.

The parts of machinery most affected are the discharge side of turbines; the suction side of pump impellers, particularly when working with a high suction lift; sharp bends in a piping system and the discharge side of regulating valves. Damage to propellers on ships is a well-known example but damage is just as likely to occur on rudders and shaft brackets. A similar effect is obtained in high frequency vibration of materials, e.g. at frequencies of 6, 500 cycles per second. This is the cause of damage to the cast-iron cylinder liners in diesel engines, which has become increasingly prevalent in recent years owing to the design of engines of lighter construction, more highly powered and working at higher speed.

A wide range of materials is susceptible to this form of attack including many metals and non-metals. On metals it would appear that this type of damage is not purely mechanical but that electrochemical action may play at least some part, the relative importance of the two depending on the particular alloy and environment. Where the conditions are particularly bad, cavitation deformation or fracture will be very severe. When the implosive forces are less and are below the yield strength of the metal the damage will be more in the nature of fatigue failure. The length of time to failure will be greater and electrochemical effects will be significant.

Finally, when the intensities of cavitation are low the electrochemical influence will predominate and result in true cavitation corrosion.

This form of attack may be minimised by the choice of less susceptible materials. Titanium, austenitic stainless steel and nickel-aluminium bronze are among the most resistant metals but are not completely immune.

Most metallic coatings are too soft and of the non-metallic coatings alumina is too brittle. The only completely successful metallic coating is nickel combined with hard chromium, but nickel phosphorus plating is partially successful. In both cases the plating must be free from pin-holes.

Resilient coatings such as rubber or nylon may be used. These not only cushion the implosive forces but also at the same time isolate the metal from the corrosive influences.

The damage is also decreased by injecting air which may act by altering the physical conditions and will also help in maintaining a protective film.

The amount of damage is increased in corrosive salt solutions but the attack can be reduced by the addition of corrosion inhibitors and often prevented by cathodic protection. Although it might be argued that the efficacy of cathodic protection is a positive indication of an electro-chemical mechanism, it is considered much more likely that the cathodic hydrogen generated at the metal surface acts as a cushion against the implosive forces.

In the case of liners in diesel engines, impulsive transverse impacts created by piston slap produce vibrations of the liner wall. The resulting cavitation can be combated by the application of hard electroplate or sprayed 18/8 stainless steel. If the rate of vibration is not too high, damage may be controlled by inhibition with 2,000 ppm sodium chromate. At a vibration frequency as high as 6,500 cycles per second, however, cast iron is not protected by even 4,000 ppm. On the Continent the corrosion has been controlled with benzoate/nitrite (pp 171, 190).

Change in piston design may also improve the position. One possible method is to increase the rigidity of the liner and another is to use an oil-cushion type of piston. The latter provides viscous damping of the transverse motion which both reduces vibration and diminishes pitting attack.

There is no sharp demarcation between cavitation-erosion and impingement attack. Small gas bubbles in the water may act as nuclei on which the cavities develop and the picture of a small gas bubble expanding and contracting with fall and rise in pressure may be applicable.

COATINGS: ALLOYING

The production and resistance of coatings is full described in Chapter X but one metallurgical aspect should perhaps be mentioned here. In all the hot-dipping or diffusion processes there is alloying of the coating with the basis metal.

Tin, zinc and aluminium all form intermetallic compounds with iron which may modify the corrosion when the outer layer is penetrated but mainly affect the mechanical properties because they are so brittle. Tin and aluminium form $FeSn_2$ and Al_3Fe respectively but the zinc coating is more complex. Next to the steel Fe_3Zn_{10} is formed and overlying this is a layer of $FeZn_7$ which is responsible for the adherence of the coating. Between this and the outer surface of zinc is the compound $FeZn_{13}$ which limits diffusion.

Lead does not alloy with iron so tin and antimony, which alloy readily with either, are added to the bath. The lead-tin solder is soft and the small amount of iron-tin compound does not prevent the coated metal from being amenable to mechanical deformation.

Chromium diffusion creates a chromium-rich surface of columnar ferrite, a narrow intermediate zone changing to austenite at about 12 per cent chromium and a ferritic core. The mechanical properties may be impaired by excessive grain-growth and carbide formation.

Chapter VIII

FLOW, TEMPERATURE AND HEAT TRANSFER

In every case there will be some relative motion between the metal and
the water. Obvious examples are ships, where the metal moves through the
water, and condensers or public water supplies, where the water moves
over the metal. In all these cases forced convection is inevitably present.
Less evident is the movement in so-called "stagnant conditions", where
currents are produced by natural convection. The influence on the corrosion rate will depend on the relative velocity, type of water and metal.

Metals are used in contact with water over a wide range of temperatures
from ambient up to the critical point, 373°C. Associated with such changes
in temperature the corrosion process may be influenced by increase in
movement produced by temperature differences and heat transfer.

The overall corrosion rate will be controlled by the rate of transport of
reactants and products to and from the surface, diffusion control, or by
the rate of reaction at the anode or cathode, chemical control. In the corrosion of metals such as iron the overall rate of attack is controlled by
the rate of diffusion of oxygen from the bulk of the water to the metal surface. With more noble metals such as copper the rate is partially controlled by diffusion of metal ions away from the surface.

There are two main types of diffusion, convective diffusion and molecular
diffusion. In convective diffusion small elements of water containing dissolved oxygen move towards the surface under the influence of natural or
forced convection currents while in molecular diffusion the individual
molecules of oxygen diffuse through the water matrix. The latter is a much
slower process than the former.

HYDRODYNAMIC CONSIDERATIONS

At low relative velocities between metal and water the flow is laminar or
streamline, small elements of water moving parallel to one another in the
direction of flow. Above a critical velocity the flow becomes turbulent and
a small element of water chosen at random may be moving in any direction, although averaged over the whole cross section the water will be
moving in the desired direction. The change from one type of flow to the
other does not occur at any precisely definable velocity but rather over a
range of velocities. Flow is dependent on such factors as surface roughness and mechanical vibrations, and is also a function of the geometry of
the system. Accordingly, for water flow in pipes the critical velocity is
so low that for most practical purposes the flow is turbulent. On the other
hand for a rotating disc the angular velocity at which the flow over the
surface becomes turbulent is quite high.

Whatever the type of flow there will be a layer of liquid in immediate contact with the metal, e.g. a pipe wall, in which the flow is laminar. This laminar boundary layer is caused by the viscous drag on the water by the metal surface. On departing from the surface succeeding layers of molecules will move with increasing velocity relative to the metal. Even when the flow is turbulent there will still be a thin laminar boundary layer (Fig. 26, A). Over this there will be a sub-layer B in which the degree of turbulence increases with distance from the surface, and finally a region C is reached in which the flow is fully turbulent. For water in pipes the two types of flow are shown in Fig. 27.

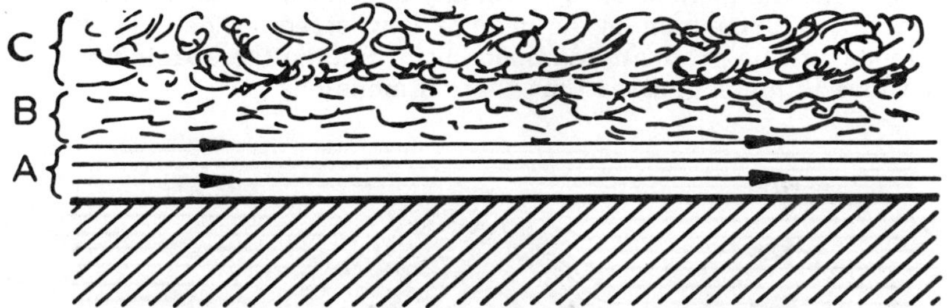

Fig. 26.   TURBULENT FLOW OVER A SOLID SURFACE

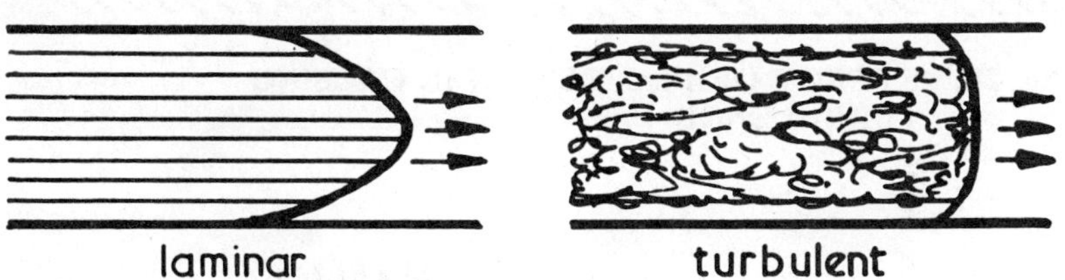

Fig. 27.   FLOW IN PIPES

At the metal-solution interface, there will be a boundary diffusion layer in which the relative velocity is small or zero and the transport of reactants will be entirely by molecular or ionic diffusion. Outside this layer the increased velocity will mean that convective diffusion will predominate. Hence the overall rate of attack in a diffusion controlled reaction will be determined by the thickness of this boundary diffusion layer and the diffusion coefficient. The thickness of the diffusion layer is about 12 to 20 mils for a stagnant solution and decreases with increasing flow to 0.5 mil or less.

FACTORS PRODUCING TURBULENCE

In practice the water may be stagnant or in laminar or turbulent flow in different parts of the same system. Turbulence will be produced or increased by a number of factors involving the lay-out and geometry of the system and will be reflected in variations in the type and rate of corrosion. These factors include badly aligned flanges, sharp changes in direction, as at pipe bends and T-pieces, and sharp changes in section as occur from a water box into condenser tubes and on the downstream side of reducing valves and pumps.

Surface roughness is a very important factor and will be caused by projections or grooves in the metal or irregular disposition of corrosion product. If the flow is laminar and the height of the projection is small in comparison with the thickness of the laminar boundary layer, the main stream of water will be unaffected. If the flow velocity is greater than a critical value the projection, as with any body of unstreamlined shape, will produce local turbulence on the rear or downstream side (Fig. 28). This turbulence will be gradually damped out providing the spacing between the projections is much greater than their dimensions.

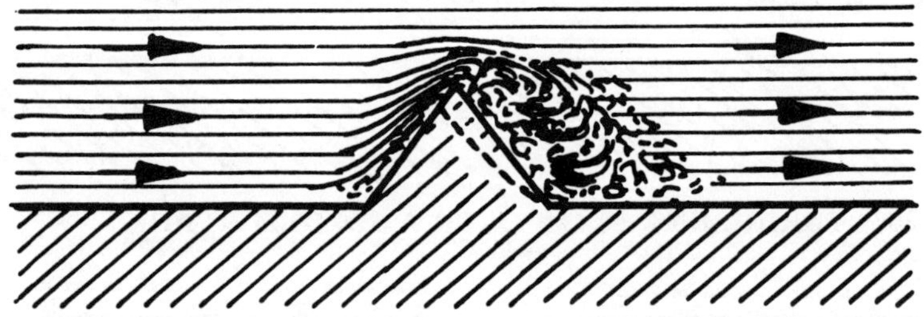

Fig. 28.   EFFECT OF PROJECTION IN CAUSING
           TURBULENCE

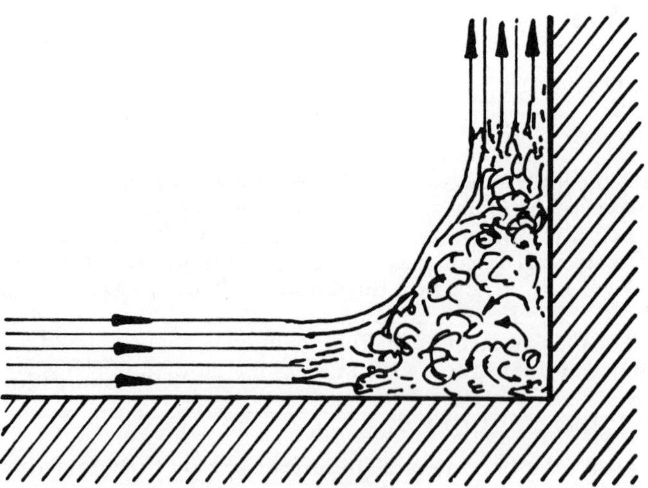

Fig. 29.   FLOW AT A CORNER

Another example is provided by a sharp change in flow direction, e.g. at
a corner (Fig. 29). Near the corner the drag on the water will produce an
increase in pressure both upstream and downstream and local turbulence.
In the case of a weir the water flow will be smooth at low velocity (Fig.
30a). If the velocity is higher the flow may still be laminar along the first
plane but will break away from the surface on reaching the edge and be-

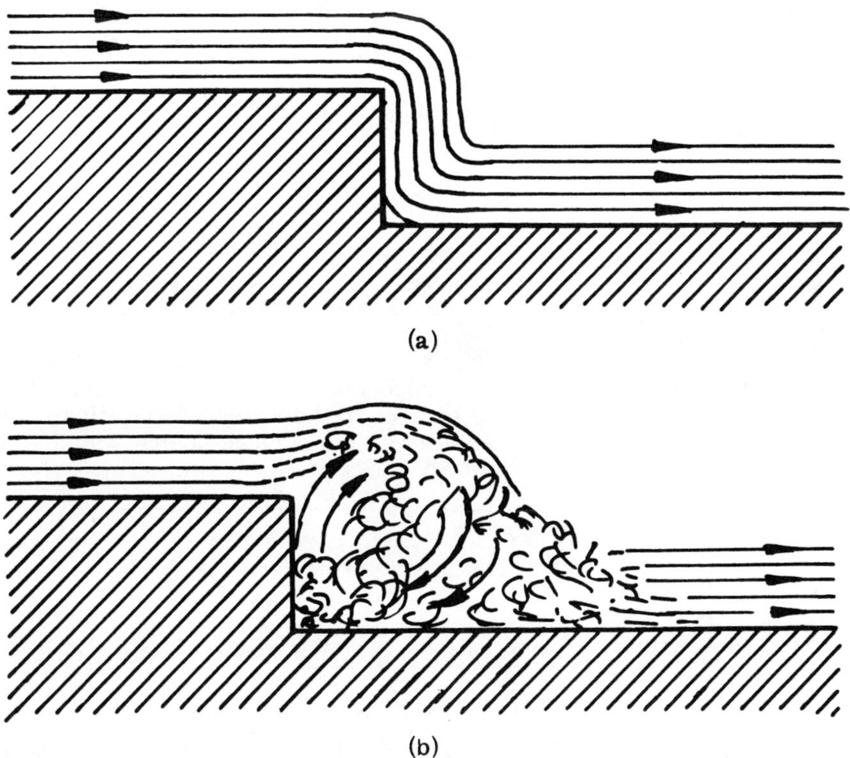

(a)

(b)

Fig. 30.   FLOW OVER A WEIR

(a) - low flow velocity, (b) - high flow velocity

come turbulent (Fig. 30b). A groove behaves as a combination of a weir
and a corner (Fig. 31). Thus any projections or corners will encourage
the production of turbulent flow close to the surface of the metal and an
improvement in the stirring of the liquid. In general, where the surface is
rough the attack on the projections will be more rapid than over the rest
of the surface. In the case of copper the increase in flow will make the
metal more anodic and the shape of the projection in Fig. 28 will take on
the form shown dotted. With metals such as iron which are easily passivated the rear of the projection will be cathodic and may produce increased attack on adjacent areas depending on the conductivity of the water.

Fig. 31.   FLOW OVER GROOVE

STAGNANT SOLUTIONS

Under idealised conditions where there is no movement, the corrosion
rate will be controlled by diffusion of reactants or products throughout
the whole bulk of the liquid and will be very low, point A (Fig. 32). In all
practical conditions convection currents exist owing to differences in temperature or density from one point to another. In the absence of significant temperature differences, such as those applied in boilers and condensers, variations in density play the preponderant role in determining the
amount of convection. At the free surface of the water, evaporation leads
to an increase in density of the solution in the surface layers, which sink
and produce movement.

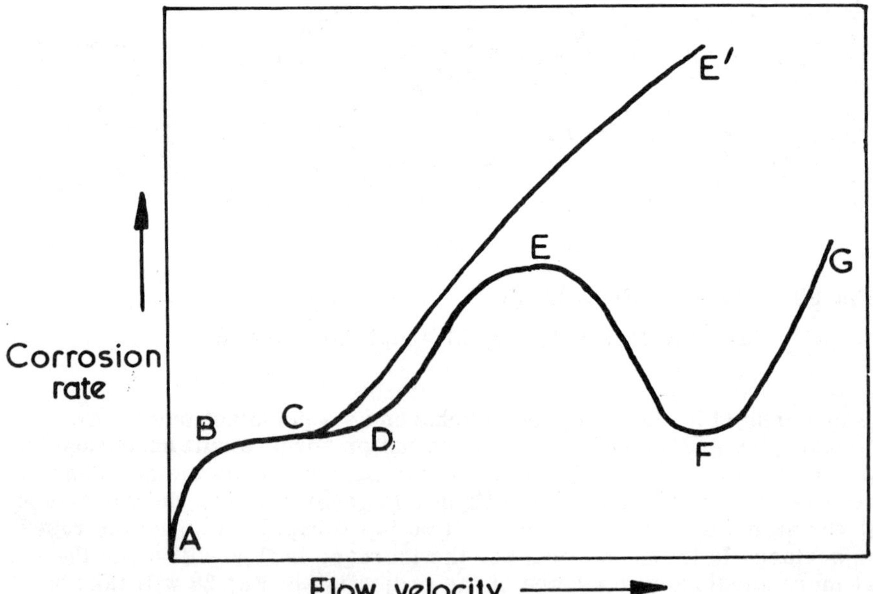

Fig. 32.   EFFECT OF FLOW VELOCITY ON CORROSION
            RATE

Natural convection will also be augmented by movement induced by corrosion. This will produce density changes which create downward or upward streaming. In many cases downward movement of the product itself will play a major part in the production of convection currents. When a sheet of iron is hung vertically in water, flow from the bottom of the specimen is observed soon after immersion and ultimately a mound of corrosion product is formed on the bottom of the vessel. On the surface of the plate the oxide film breaks down locally and movement under gravity leads to the formation of vertical streaks of product which themselves set up oxygen concentration cells in two ways. Firstly, oxidation of the downward flowing product leads to a lower oxygen concentration on parts of the metal and, secondly, the corrosion product itself will restrict the access of oxygen to the surface. In a similar manner the downward flow of cathodic alkali will provide a degree of protection on other areas. As a result a relatively static pattern of attack develops on the metal surface, in the case of iron, streaks of rust, within which there is usually a number of primary and secondary pits, and an unattacked area. In many cases it is clear that the attacked area is cathodically protecting the unaffected region.

Under so-called "stagnant" conditions the measured corrosion rate is nearer to point B than point A (Fig. 32).

LAMINAR FLOW

A certain amount of movement is to good advantage. Any of the gravity effects outlined already are eliminated once the flow velocity exceeds a certain figure and the tendency to localised attack is reduced. The overall corrosion rate is not increased to any great extent (BC, Fig. 32). Under flow conditions the rate of attack, k, may be expressed as a function of velocity, v, in the form:

$$k = K_O \cdot v^n$$

where K_O and n are constants. For water flow in a pipe and for spheres falling through a liquid, n is one-third, while for a rotating disc or a plate placed in a current of water parallel to the direction of flow, n is one-half.

TURBULENT FLOW

The variation in the corrosion rate in turbulent flow is more dependent on the type of water. For a relatively pure water, free from aggressive ions, a curve of the general shape DEF (Fig. 32) will be obtained for a number of materials. Up to the point E the corrosion rate increases ever more rapidly with the amount of turbulence until ultimately the rate of attack is proportional to the flow velocity, i.e. n is unity. In many cases, however, once a certain oxygen supply to the surface has been attained the corrosion rate begins to decrease along the line EF. This can be produced by a change in the nature of the corrosion product which becomes more continuous and adheres effectively to the surface thereby producing a physical barrier against diffusion of corrosive reactants. In other cases the ability for the metal to passivate is increased by the increased supply of oxygen. Ultimately the metal may become passive and no observable corrosion will then take place.

With further increase in flow velocity corrosion-erosion may be encountered and the attack will increase once again, FG. Here both impingement and cavitation attack become more serious with increasing flow velocity (Chap. VII).

In waters which are more aggressive, the rate increases along the line CE' with increase in flow velocity and in brackish waters the corrosion rate of iron may be as high as 2000 mdd at a velocity in the region of 7 ft sec.

This is, of course, a very generalised picture and Fig. 32 would differ according to the metal and solution. Nevertheless, the influence of the concentration of aggressive ions is more marked at higher flow rates. At the same time as the access of the aggressive ions to the surface is increased, the supply of inhibitive ions and scale-forming ions is also increased and the corrosion may be stifled if the second effect is dominant. The degree of turbulence will be increased by the factors already mentioned and will have the effect of displacing the curve FG to the left.

TEMPERATURE

Water is used in contact with metals over the whole range of its liquid phase stability, from the freezing point, 0°C, up to the critical point, 374°C, at which temperature the liquid and vapour phase coalesce. Changes in temperature in the lower range from, say, 0 to 30°C are produced by variations in the source of the water and the time of the year. Above, and overlapping with this range, is water used for cooling purposes. At still higher temperatures, 100 to the region of 120°C, water is used for steam generation, for industrial processes and, to a less extent, for small power generators. In the highest range, up to and beyond the critical point, steam is raised in boilers for power generation.

Throughout these ranges, increase in temperature affects the chemical composition and physical properties of waters, the nature and properties of deposits and the actual behaviour of the metal itself. The water composition is affected by changes in the stability and solubility of the dissolved solids. Calcium bicarbonate is decomposed to calcium carbonate, sodium carbonate is hydrolysed to caustic soda with loss of carbon dioxide, and magnesium salts are hydrolysed to give solutions of low pH values. The solubility and diffusion rate of dissolved gases also vary appreciably with temperature and of particular interest in corrosion processes is the behaviour of oxygen.

The change in composition of the corrosion product may affect the temperature dependence of corrosion rate as in the case of zinc. Again, in the case of iron the corrosion products at ambient temperature may include magnetite and hydrated ferric oxides but at 200°C and above, only magnetite and the anhydrous ferric oxides are stable.

The physical properties of the metals themselves are also changed, e.g. copper alloys are unsuitable for use under boiler conditions owing to decrease in strength. Thermogalvanic effects arise in which the potential of the metal varies with the temperature. Thus, in neutral chloride solutions an increase in temperature in iron leads to a decrease in potential, i.e. hot part anodic, and it follows that temperature differences can cause enhanced corrosion on certain parts of a structure.

In heat-exchange equipment, such as condensers and boilers, heat has to be transferred across the metal surface. In order for this to take place the temperature on the metal surface must be higher than that in the bulk of the liquid and, hence, this is the very site on which conditions are more favourable for the deposition of calcium carbonate and sulphate scales. Although such a layer may effectively stifle the corrosion reaction it may seriously interfere with the heat transfer. This will lead to still higher metal temperatures and possibly rapid oxidation and failure of the metal— "overheating" or "burning".

TEMPERATURES NEAR AMBIENT

In warm and hot water the tendency to pitting shows an appreciable increase. It is generally known that heat-exchange equipment operating at too high a temperature may lead to premature failure.

During the summer months the higher temperatures of the surface waters used in cooling, e.g. sea water, can produce more rapid corrosion. New condenser tubes have been known to give a longer life when installed during the winter than in the summer, due to the difference in the type of protective film formed. On copper-zinc alloys the corrosion product formed in cold water is generally thinner and more continuous than one formed in warm water.

Fouling and pollution are also greater in the summer than in the winter because the higher temperature favours the breeding of marine organisms and the decomposition of vegetation. It follows that in northern waters there is more fouling in summer than in winter although in the tropics fouling is almost continuous throughout the year.

In sea water the rapid corrosion of brass condenser tubes and cast bronze impellers during the summer season was attributed to the presence of hydrogen sulphide at concentrations up to 15 ppm. Similar effects may be found in fresh waters which do not deposit protective scale.

OPEN SYSTEMS

In the case of equipment open to the atmosphere the temperatures range from 0 to 100°C. In this region the solubility of oxygen decreases continuously to become zero at the boiling point, although complete exclusion of oxygen is possible only if certain precautions are taken. On these grounds it would be expected that corrosion reactions under cathodic control would decrease with increase of temperature. However, over the same range of temperature the rate of diffusion of oxygen will be continuously <u>increasing</u>. The net result is that the corrosion of a number of metals, in which the reduction of oxygen is the main cathodic reaction, have a maximum in corrosion rate in the range 70 to 80°C. Iron behaves in this manner, the rate increasing up to 80°C under the influence of increasing diffusion rate and then decreasing up to 100°C under the influence of the reduction of oxygen solubility (Fig. 33).

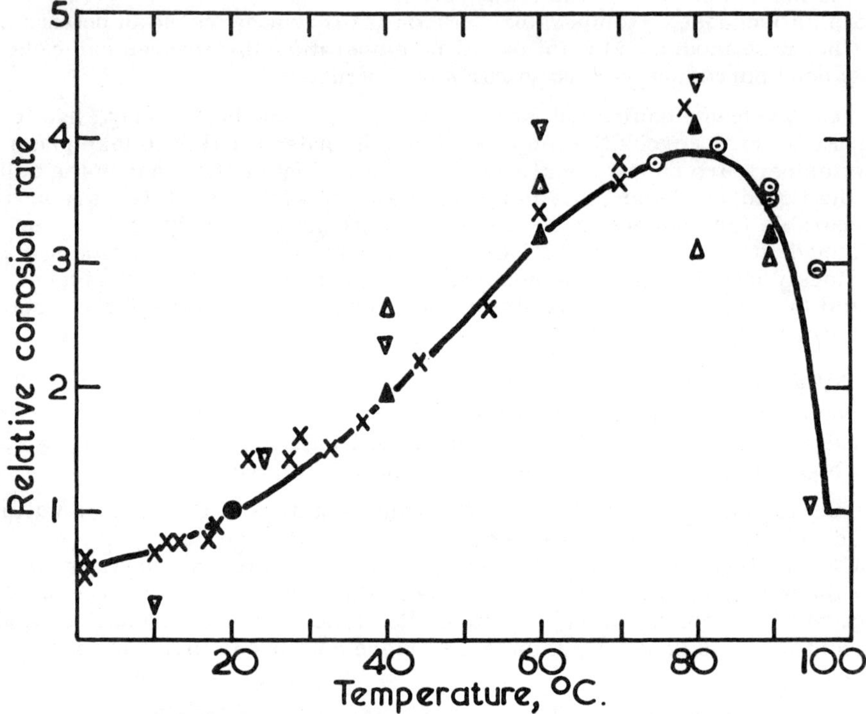

Fig. 33 EFFECT OF TEMPERATURE ON CORROSION RATE IN AN OPEN SYSTEM

 △ - Heyn and Bauer, distilled water, 22 days.
 ▲ - Heyn and Bauer, intermittent heating.
 ▽ - Rice, Pittsburgh water, 8 hours.
 ×
 ⊙ - Friend, distilled water.

At lower temperatures in dilute chloride solution the corrosion rate of mild steel approximately doubles for a temperature rise of 30°C. In the range 0 to 75°C in 3.0 per cent sodium chloride the corrosion rate of Admiralty brass and cupro-nickel doubles for a 20°C rise in temperature.

The rate of pitting of stainless steel in strong chloride (4.0 to 10.0 per cent sodium chloride) increases with temperature and the maximum weight loss due to pitting occurs at 90°C. In more dilute solutions the maximum occurs at a still higher temperature. At the boiling point the oxygen is expelled and there is no pitting irrespective of the chloride concentration.

In the case of zinc the corrosion rate is related to the change in the nature of the corrosion product. From 20°C to 50°C the rate approximately

trebles (Fig. 34) and the product becomes less gelatinous and its adhesion
is reduced. In the range 50 to 65°C there is a marked increase in rate, the
product becomes decidedly granular, flaky and non-adherent. In the range
70 to 85°C this type of product is accompanied by marked pitting. At
higher temperatures the rate decreases again owing to the decreased oxygen solubility and the product becomes compact and dense and increases
in adhesion.

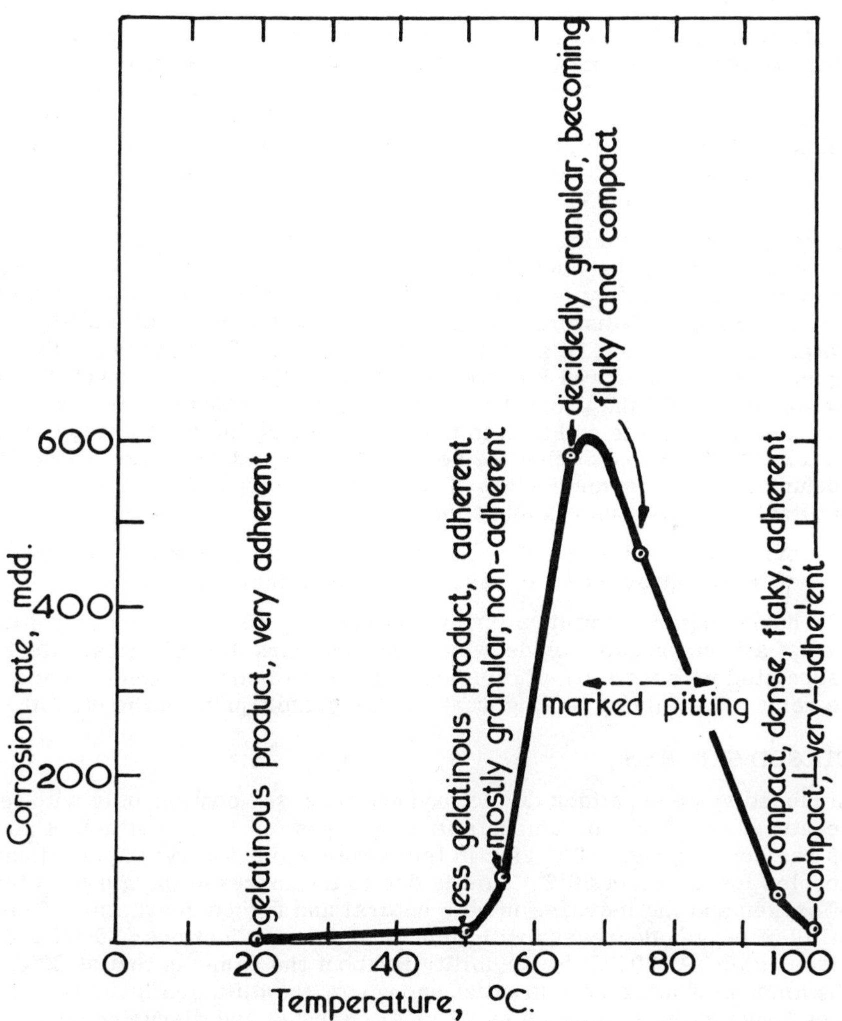

Fig. 34. INFLUENCE OF TEMPERATURE ON THE
CORROSION OF ZINC

(after Cox, G.L., Ind. Eng. Chem., 23, 902, 1931)

The attack is reduced by carbonate hardness. In closed hot-water systems highly localised pitting often occurs with soft water but this is less likely in hard water.

In soft waters with appreciable bicarbonate content the possibility of a potential reversal of the zinc coating on galvanised iron leading to accelerated attack on the basis metal should be borne in mind. Waters with a higher chloride content are more aggressive to zinc and the likelihood of a potential reversal is reduced.

In strong chloride solution the attack on pure magnesium is relatively constant over the range 20 to 100°C. Magnesium alloys show a tendency to deep pitting with increase of temperature if cathodic impurities such as iron, nickel and copper are present above the tolerance limit. In strong chloride there is a 30-fold increase in rate on increasing the temperature from 17 to 55°C. This corresponds to a rate increase factor of about 2.3 per temperature rise of 10°C.

Although with increasing temperature the corrosion rate of aluminium shows a maximum between 70 and 80°C, the rate is, in fact, low and aluminium is used widely in contact with natural waters and is particularly useful for steam radiators and the storage and transport of distilled water. As in the case of zinc, the corrosion rate of aluminium with increasing temperature is largely dependent on the type of product formed. Below about 70°C the product first formed is amorphous which eventually changes to boehmite and then to bayerite. During the first change the attack increases and during the second the attack decreases. Above 70°C, boehmite only is formed on top of the initial amorphous film and the rate of attack falls with increasing time.

The oxide film is destroyed by alkali and certain waters and also the presence of copper ions can cause localised attack and pitting.

The darkening of aluminium cooking utensils is connected with the formation of a film, probably oxide, which appears dark due to optical effects associated with its state of division. This film is quite harmless and can be removed easily by boiling weak acid, e.g. fruit juice in the utensil.

CLOSED SYSTEMS

In closed systems, attack due to oxygen increases continuously with temperature. On steel this amounts to an increase in rate of attack of about 25 per cent for each 10°C rise in temperature, i.e. the rate approximately doubles for a rise of 30°C. This is due to the increase in rate of diffusion of oxygen and the increase in both natural and forced convection. The solubility of oxygen decreases with temperature up to just over 100°C and then rises again. At 200°C the solubility is about the same as that at 20°C. If the amount of movement is great enough, or the diffusion layer on the metal surface is influenced by thermal currents and disrupted by evolution of steam bubbles, the rate of attack may no longer be controlled by diffusion of oxygen to the surface but by the rate of reaction at the surface. In this case we can expect a much higher increase in rate with temperature i.e. up to double for a rise in temperature of 10°C. This may be reduced by the deposition of a calcareous scale on the surface but this will also interfere with the heat transfer.

In the absence of oxygen the direct reaction between iron and water is theoretically possible at ordinary temperatures but is not observed in practice. In pure water or very dilute solutions gas bubbles only appear on steel when the temperature is about 95°C but on cast iron the evolution of hydrogen is rapid at temperatures from 60 to 80°C and is measurable at 40 to 45°C. The product of pure iron in oxygen-free water is ferrous hydroxide but at 60°C magnetite is formed and there is a diminution in the amount of dissolved iron.

HEAT TRANSFER AND CORROSION

Knowledge of the complicated problem of the interaction between heat transfer and corrosion is fragmentary. Nevertheless, from practical experience with heat exchangers and boilers, it is clear that a much higher rate of attack, and even an entirely different type of attack, can occur on a surface across which heat is being transferred. In acid solutions it has been found that the corrosion rate of a metal surface transmitting heat may be as high as 12 times that for the metal immersed in the solution at the same temperature. Such information indicates that the skin temperature of the metal is the main controlling factor—not the amount of heat flowing through the metal.

The way in which this heat transfer takes place can be illustrated by the example of water being heated by a horizontal metal wire. For heat to be transferred a temperature difference, Δt, must exist between the surface of the wire and the saturation temperature of the bulk of the liquid and this difference will depend on the heat flux Q (Fig. 35). In the range from ambient temperature, A, to the boiling point, B, heat is transferred through a thin superheated layer of liquid on the metal surface in which the temperature falls from the metal temperature to that of the bulk liquid, the difference varying from a very small value up to about 10°C depending on the rate of heat transfer. In the bulk of the water heat transfer takes place by convention currents produced by the temperature differences and density differences, enhanced by evaporation from the free surface of the water. Convection currents will also be increased by the disengagement of dissolved gases, which can produce corrosion problems in water heaters owing to the <u>hot-wall effect</u> (discussed later).

At a certain point B, when the rate of heat transfer has attained a high enough value, bubbles of vapour will start to form at discrete points on the metal surface and the liquid will start to boil. After breaking away from the metal surface these bubbles rise through the liquid producing a stirring effect, which augments the convection currents already existing. As the temperature of the heated surface is increased, more heat is transferred and the rate of boiling increases until a point C is reached when the bubbles formed on the heated surface have become so numerous that they coalesce to form a continuous sheet of water vapour over the whole surface. Having a low thermal conductivity this vapour film will act as an

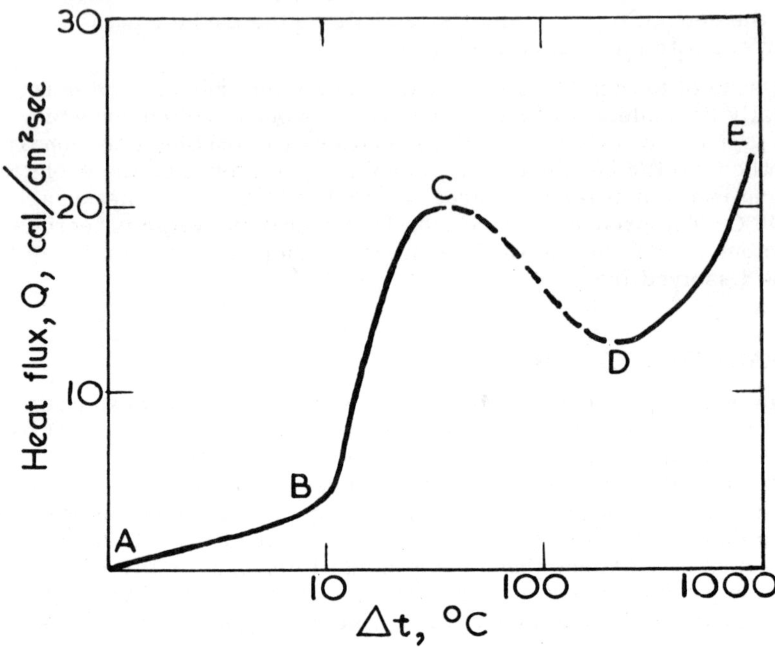

Fig. 35. HEAT TRANSFER CURVE FOR HOT WIRE IN WATER.

(Nukiyama. S., 1934. J. Soc. Mech. Engrs. Japan. **37**. 367-**74**, 553-4)

insulator to the flow of heat and in the range CD the heat flux decreases with increasing Δt. Subsequently in the region DE the heat is transferred through the vapour film by conduction and convection and <u>film boiling</u> occurs. Very high values of Δt are reached in this region leading ultimately to overheating and then melting of the metal. Here conditions are favourable for the direct reaction between metal and steam.

In boilers the important region is BC where <u>nucleate boiling</u> occurs. At a given temperature the vapour pressure over the very small concave surface of a steam bubble is less than that over a flat liquid surface. Hence the liquid must be hotter to evaporate into a small bubble of vapour than into the vapour space above the liquid and the difficulty in forming vapour bubbles is apparent. Thus bubbles form preferentially on active nuclei on the heating surface where the temperature and the nature of the surface are favourable. On some surfaces the vapour bubbles form readily and quickly detach themselves whereas on others they form only when the liquid has a higher degree of superheat. At high rates of heat transfer this can amount to 20°C.

On a rough surface there is a larger number of nuclei than on a smooth one and the liquid boils with about half as much superheat. At low heat flux, since a substantial portion of the metal is covered with liquid, a considerable fraction of the heat transferred is first used to superheat liquid. As a bubble leaves the heating surface and rises through the water its size increases since superheat is consumed in vaporising additional liquid. The increase in volume ranges from 4, 500 times for small bubbles to 140 for large ones.

A most important property of the surface is the angle of its contact with the liquid. If the surface is lyophilic, the contact is greater than 90° and tends to drive the bubble off the surface. If the surface is lyophobic, e.g. covered with oil, with a contact angle of less than 90° the bubbles tend to spread out over it and consequently tend to promote film boiling at a comparatively low rate of heat transfer. (Fig. 36).

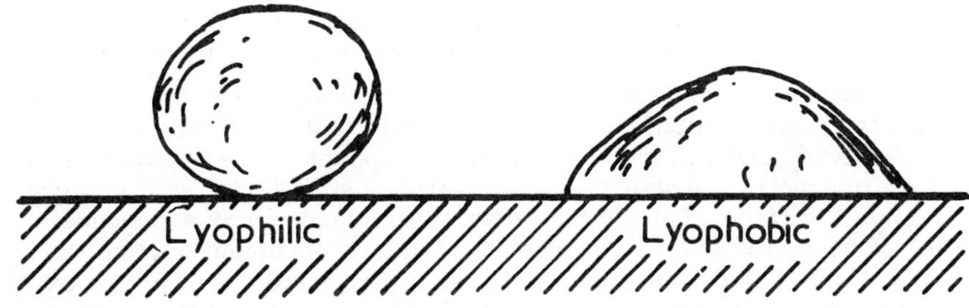

Fig. 36. LYOPHILIC AND LYOPHOBIC BUBBLES

Any contamination of the surface that affects its wettability influences the size and shape of the vapour bubbles, which in turn have an effect on the heat transfer. The effects of addition agents, scale and corrosion deposits and artificial roughening of the surface can be significant in nucleate boiling but no satisfactory general correlation has yet been made.

Most of the experimental work on heat transfer phenomena has been done with clean metal surfaces. Relatively small amounts of corrosion will modify the surface to an extent sufficient to influence the heat transfer significantly by providing nuclei for bubble formation.

In nucleate boiling the turbulence produced at the immediate metal surface by the steam bubbles will modify considerably material transport to and from the surface. At low heat flux this disturbance will occur at a limited number of points on the surface although the increase in convection produced by heat transfer at other areas will also influence material transport.

In bubble formation both the excess temperature at the point of formation and the evaporation of water into the bubble near the metal surface can affect corrosion. The excess temperature may well modify the corrosion potential of the metal at the site of bubble formation, giving rise to a thermogalvanic cell:

metal at temperature t_1/solution/metal at temperature t_2

where t_2 differs from t_1. It has been shown that for copper, nickel, iron and lead in acid, neutral or alkaline media the hotter metal is always the anode. In view of the possibility of high local superheat some association between pitting and bubble formation may be expected. Hemispherical and horse-shoe shaped pits from their shape may be linked to bubble formation at the site of attack (Pl. 19). Crevice attack can take place between the bubble and the metal, and the question whether a hemispherical or horse-shoe shaped pit is formed depends on the size of the bubble. Production of corrosion product at nuclei may assist in maintaining their activity. In a solution boiling at ordinary pressures it is readily observed that bubbles continually form from the same nuclei and often have the appearance of a stream of gas emerging from holes in the metal. Thus conditions at active points may persist over long periods of time.

Steam bubbles forming at isolated points may affect the corrosion process in another way. Evaporation into the bubble will result in a concentration of dissolved solids in the water in the periphery of the bubble and, even at a bulk concentration well below saturation, deposition of salt can occur on the metal surface. The nucleus may, therefore, become surrounded by a ring of deposited salt. Even when the nucleus becomes inactive the salt may influence the corrosion behaviour by, for example, affecting oxygen or salt transport to this area and producing a concentration cell. Even if the salt redissolves after a time its effect will be continued by the corrosion product.

These tentative ideas on the interaction between bubble formation and corrosion suggest that the rate of pitting may be reduced, even if the total corrosion remains the same, if an adequate number of nuclei for bubble formation are initially available on boiler tubes. A weathered surface with some mill scale may be more suitable than a pickled surface on which all the chemically active spots, which may also be active nuclei for steam bubble formation, have been removed. Clearly an examination of the relation between the two types of nuclei and an investigation of the interaction between nucleate boiling and pitting would repay further study.

Another factor influencing the attack on heat transfer surfaces is the formation of carbonate scales. Service tubes covered with thick scale layers often suffer attack beneath the scale, which is disrupted by the rust formed. If the thermal-insulating scale layer is thick enough, the temperature at the metal surface may rise sufficiently high for water penetrating the scale to react, as steam, with the iron surface at an appreciable rate, the magnetite formed disrupting the scale. Although the steam-iron reaction is only appreciable above 410°C, if the metal is in a more active form, appreciable attack may take place at a temperature even as low as 230°C. On the other hand, other metals, such as copper, may catalyse the decomposition of steam at lower temperatures, the oxygen formed reacting with the iron.

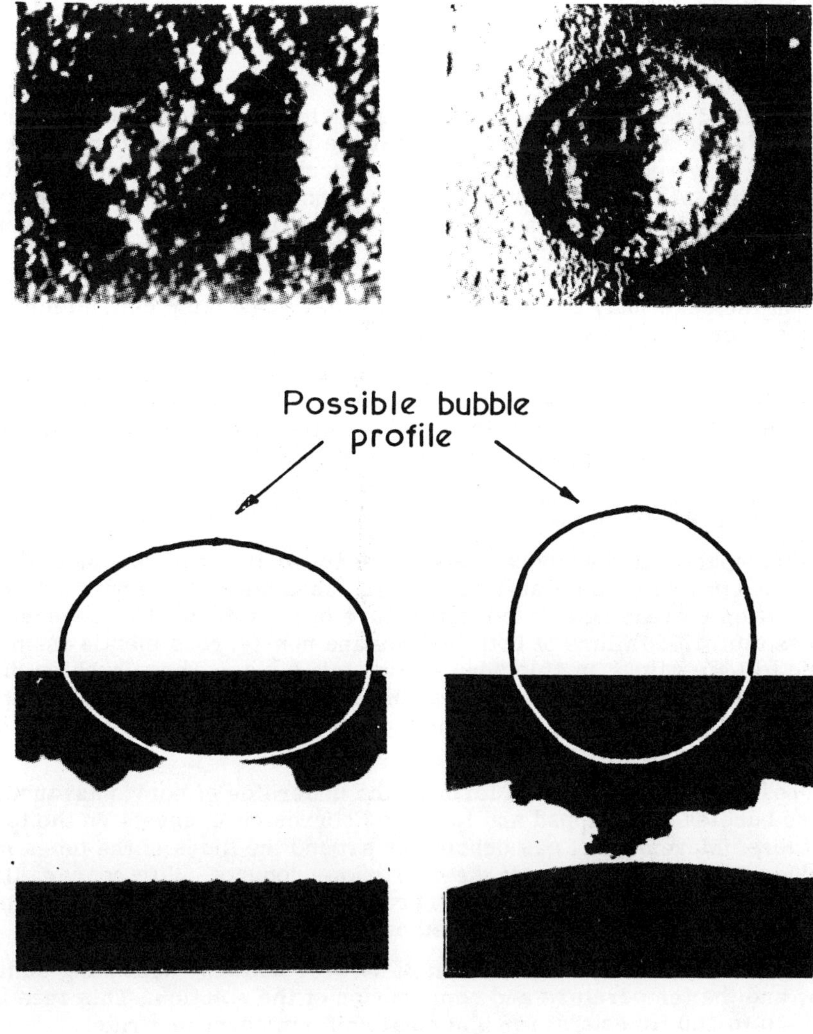

Plate 19. PLAN AND SECTION OF (left) HORSE-SHOE PITS AND (right) SPHERICAL PITS.

The value of having any deposit at all on the metal surface, except possibly a very thin continuous film, is questionable. The lime-soda treatment, which leads to the deposition of such scales, is so convenient a method of treatment, however, that its retention is likely, particularly in low pressure boilers.

If the scale deposition is localised, deep pitting may occur along the edge of the affected areas. It is possible that the incorporation of magnetite in the scale may render it cathodic to the surrounding metal inducing pitting of the latter. Alternatively attack may be produced by the low <u>thermal</u> conductivity of the scale resulting in the metal beneath the scale having a slightly higher temperature and causing a flow of heat from the scale-covered metal. The slightly higher temperature of the metal along the edge of the scale may then give rise to a thermogalvanic cell, attack taking place preferentially on the hotter metal.

The same mechanism may apply to attack at the base of pores in less compact scales. The oxygen supply in the pores will also be restricted and a differential aeration cell will form between the base of the pores (anode) and the surface with free oxygen access (cathode).

HOT-WALL EFFECT

In water heaters and evaporators with steam or flame on one side of the tube, deep pitting on the water side of the tubes takes place with a rate of penetration increasing as the temperature of the tube wall increases. This causes the rapid failure of both ferrous and non-ferrous metals when exposed to waters high in chloride. Pitting takes place where bubbles of gas repeatedly separate from water, usually at scratches and sharp indentations, and the bubbles of gas clinging to the heated metal prevent cooling and give rise to <u>hot spots</u>.

The resultant deep pits often form on the underside of horizontal tubes where bubbles are trapped and localised corrosion is absent on the top of the tubes. Movement of gas bubbles up around the sides of the tubes may produce upward streaming of the corrosion products. With copper alloys nodules of black or greenish-black products, mainly cupric oxide, form over deep pits filled with bright red crystalline cuprous oxide.

The rates of penetration vary from 25 to over 250 mpy depending on the alloy and the temperature and composition of the solution. This rate is from 10 to 100 times that for a normal uniform corrosion rate.

The important factors are the local high temperature, the concentration of corrosive reagents separating out where the bubbles form, the composition of the gases and, in particular, the presence of depolarising gases.

In order to avoid or reduce this form of attack it is necessary to keep the operating temperature as low as possible and to avoid continuous uninterrupted heating. The dissolved gases or air should be reduced as far as practicable and it is obviously useful to ensure that as many nuclei as possible are available as alternative points to which the formation of bubbles will be diverted.

COLD-WALL EFFECT

Condensed droplets can absorb small amounts of aggressive materials such as hydrochloric acid, ammonium hydroxide and sulphur dioxide which may lead to deep grooves on the sides of the tubes. The droplets tend to form at the same spots repeatedly and roll down the same area. Longitudinal grooves may also form or pitting may occur on the bottom of tubes where the droplets hang for some time before dropping off. Much more thinning of a tube wall usually takes place on the bottom of the tube than on the top.

BOILERS AND HIGH TEMPERATURE PLANT

Modern boiler technology covers the temperature range from 100°C to over 374°C. The internal pressure will range from 100 to 5000 lb per sq in and the final steam temperature from the saturation level to 650°C. The quality of the water used varies with the temperature and pressure. At the lower end of the scale relatively brackish waters, high in dissolved solids, may be tolerated. At the upper end not only is it essential to have the purest water possible and complete absence of oxygen but water treatment must not introduce undesirable solids (Chap IX).

For boiler applications mild steel has excellent mechanical properties, which are retained up to relatively high temperatures, but its poor corrosion behaviour requires careful control of the water quality. Above 450°C small percentages of molybdenum or of chromium and molybdenum are effective. At still higher temperatures it is necessary to use austenitic stainless steels.

In the atomic energy field, the poor corrosion properties of iron have led to the use of other materials with higher corrosion resistances since cost is less important than the avoidance of radioactive hazards. In water containing 1.4 ppm oxygen, titanium, zirconium, hafnium, platinum, austenitic stainless steels and certain cobalt alloys are unchanged except for a tarnish film. On the other hand nickel is only resistant up to 200°C and copper and aluminium are limited to temperatures below 150°C under such conditions.

Stainless steel and cobalt alloys undergo severe attack even at 260°C in crevices, e.g. between rivet heads and plates, with oxygen concentration in the range 7 to 15 ppm. This effect is not aggravated by dissimilar metal contacts but may be reduced if a clearance of 5 mils is maintained.

ALUMINIUM

Aluminium and its alloys are relatively resistant to corrosion by water at temperatures up to 350°C. The metal surface becomes covered with a thin adherent oxide layer which is often blistered, particularly above 200°C, by internal hydrogen evolution. The accompanying intergranular attack can be changed to general corrosion by heavy cold working.

Improvement of the corrosion behaviour of aluminium by alloying is largely dependent on achieving an even distribution of the cathodic zones on which hydrogen discharge takes place. Thus the addition of small amounts of nickel is beneficial presumably because the ease of liberation of hydrogen on nickel, i.e. low overvoltage, favours the liberation of molecular hydrogen and the inward diffusion of atomic hydrogen is avoided. The beneficial action of nickel is assisted by iron. The nickel should not be less than 0.5 per cent and the sum of the iron and nickel contents must be at least 0.8 per cent. This behaviour is associated with the formation of the intermetallic compound $FeNiAl_9$ and although numerous other alloys have been studied the iron-nickel-aluminium alloys have the highest corrosion stability.

The adverse effect of silicon can be counteracted by increasing the nickel content. An alloy with 9 per cent silicon and 1 per cent nickel and one containing up to 3 per cent silicon, 0.1 per cent copper, 1 per cent iron and 0.5 per cent nickel have behaved well in reactors at 275 and 350°C.

Aluminium alloys should not be used in alkaline water since the amphoteric character of the oxide layers limits the safe pH range to between 5.0 and 7.0.

IRON

The corrosion of iron and steel at high temperatures is dependent on the water composition, its oxygen content and the condition of the metal surface. In the "as received" condition the tubes for boilers are usually covered with an imperfect layer of mill scale. This scale, which is mainly magnetite, would be protective if it were complete and, in the treatment of boiler waters, the main object is to provide conditions where such a thin layer of magnetite is formed and kept in a state of continuous repair. In low pressure boilers, however, where the control of the boiler water is not all that would be desirable, pitting is liable to occur at breaks in this original mill scale. Such damage is caused by weathering, handling during bending or straightening, or where the tubes have been expanded into the tube plates. The pitting often occurs at the intersection of "luders" lines and, in soft water, penetration can produce perforation of new tubes in a matter of a few weeks. This localisation is due to the large cathode, magnetite-covered area, and small anode, breaks in the scale where the metal is exposed.

The action of water on iron in the absence of oxygen leads to the formation of ferrous hydroxide which is converted into magnetite or, at higher temperature, magnetite is formed directly, both reactions being accompanied by hydrogen evolution. In the presence of oxygen, however, hydrated ferric oxides are formed at all temperatures up to about 160°C above which the anhydrous forms appear.

Magnetite can form a protective film on the iron surface under favourable conditions but mill scale, rust, grease and oil will interfere with its continuity and protectiveness. This film can grow to quite appreciable thickness and may reach 10 mils on steel which has been exposed to boiler water for 10 years at 1500 lb per sq in. The growth of a protective magnetite film is accelerated by a rise in temperature and alkalinity but is interrupted by ingress of oxygen in the make up or access of air when the boiler load is lowered.

If the caustic soda is sufficiently concentrated it will attack magnetite with the formation of soluble ferroates and ferrites, i.e. Na_2FeO_2 and $NaFeO_2$. The inordinate growth of magnetite at pressures greater than about 500 lb per sq in can also give rise to trouble on steam generator tubes and above 1000 lb per sq in this is accompanied by permanent embrittlement of the underlying steel leading to sudden failure. This is probably connected with evaporation and heat transfer. Part of the hydrogen liberated in the reaction diffuses through the nearby steel, reduces carbide and caused intergranular attack.

The formation of a protective film of magnetite may be aided by high pH and low oxygen content. However, if the water becomes acid, e.g. due to the hydrolysis of magnesium salts present, the formation of a protective film of magnetite is assisted by the presence of about 0.25 per cent copper in the steel which catalyses the formation of magnetite from ferrous hydroxide.

In locomotive boilers, which have copper end plates, serious galvanic effects might be expected but in practice little trouble is experienced since the area of steel is much greater than the area of copper and the oxygen is low. Troubles due to grooving of tubes close to copper tube plates is generally due to distortion of the tubes where they have been altered for entry into the tube plates or is caused by mill scale hammered into tubes.

In recent years the practice of cleaning the internal surface of boilers by pickling with inhibited acid has been gaining favour. Even if the whole boiler is not treated in this way it is advisable to treat the tubes by shot- or grit-blasting. This provides better conditions for the subsequent formation of a coherent continuous film of magnetite with a uniform thickness.

The adhesion between any protective layer on the metal surface can by destroyed by mechanical or thermal shocks, the latter being more likely in high pressure boilers. Care should be taken in overhauling boilers that the protected surfaces are not subjected to harsh hammer blows and similar mechanical damage.

As far as possible stress should also be avoided in boilers because it is one of the factors involved in stress corrosion and caustic cracking. The introduction of welded tubes has eliminated the corrosion risk of residual stress at the ends of expanded tubes.

In high temperature boilers there is a danger of cracking through sudden changes in temperature. This is made more serious by the use of austenitic steels which have a lower thermal conductivity and higher expansion coefficient than ferritic steels. A rise in the temperature of the steam of 15°C per minute on starting and a fall of 2.6°C per minute on stopping is considered safe. Even in the absence of salt, stress corrosion cracking of the transgranular type can be produced by rapid change of temperature.

Thermal stresses also arise due to the temperature gradient through the tube walls which is greater at higher heat transfer rates and in thicker tubes. Local variations can also occur owing to the poorly adjusted heat release from the furnace or from poor water circulation at sharp bends or tube constrictions.

It would be unreasonable to expect that all boilers would be run under the best conditions, if not on the grounds of the absence of skilled personnel at least on the grounds of cost. There must be a large number of boilers in which the raw waters are not treated in any way and which operate satisfactorily even if under such conditions the steam generation will be generally far from efficient. Scales, consisting of calcium carbonate and magnesium hydroxide, and up to 200 mil thick have been observed by the authors in trawlers running on sea water. In all marine boilers there is a danger of leakage of sea water into the boiler water, usually from the condenser, and this can give rise to "scab pitting" at breaks in the scale during operation and to "soft scab pitting" during shut down. The latter form of attack occurs mainly in superheaters and headers wetted by water primed from the boiler.

Hide-out due to the deposition of sodium compounds can be a serious matter in marine boilers. Additions of lignin waste and quebracho tannin can be used with advantage at pressures up to 850 lb per sq in at rates of heat transfer below 85, 000 B. Th. U. per sq ft.

Air bubble pitting in which bubbles of air collect beneath the roof of the drum can produce intense attack. Once the oxygen is used up the nitrogen remaining forms a screen which can give rise to a differential aeration if the water contains oxygen.

Although doubly-distilled water may be used, troubles can occur in once-through boilers due to salt deposits. Molten salts deposited on a hot tube can be fluxed and, if the deposit is alkaline, sodium ferrite may be formed directly or from deposits of sulphate or chloride from which hydrochloric acid or sulphur trioxide have been expelled.

The results of corrosion are many and various. A small perforation of a boiler or superheater tube of the pinhole type may leak for a considerable time before it is found. On the other hand more general attack and wastage can give rise to a sudden opening out of a boiler tube over many inches with a rapid loss of boiler pressure. Where the failure is sudden and spectacular the whole boiler may explode, devastating buildings and causing loss of life. Such occurrences are rare but can happen with both low and high pressure boilers and management should have such possibilities always in mind when dealing with any steam-raising equipment.

MAGNESIUM AND ITS ALLOYS

Magnesium and its alloys are used for containers of atomic fuel elements since they are cheap and have a low adsorption cross section for slow neutrons, but their thermal conductivity is only 65 per cent of that of aluminium. Their corrosion resistance depends on the water composition, particularly the pH value and oxygen content. The optimum pH value is about 10 and the corrosion rate is 50 times greater at a value of 8.9 than at 9.7. A magnesium alloy with 9 per cent aluminium, 2 per cent zinc and 0.1 per cent manganese has a comparatively low corrosion rate.

ZIRCONIUM

In pure water at high temperature a protective oxide film of dioxide, ZrO_2, is formed on zirconium and at first thickens at a rate proportional to the third or fourth root of the time. Subsequently it thickens at a rate proportional to time when a non-adherent product is formed and the protective oxide film is damaged by spalling. This effect is increased by impurities in the zirconium such as nitrogen, hydrogen, carbon, titanium and silicon. In general, the rate of attack on the metal surface at temperatures up to 400°C is only 1.8 to 3.6 mdd.

The properties of the metal may be improved by alloying with small amounts of tin, the best known alloy being Zircaloy 2 with 1.5 per cent tin and small amounts of iron, nickel and chromium. This alloy has other advantages besides its good corrosion stability in that it is not necessary to have very pure zirconium for its preparation. The addition of tin improves its mechanical strength and the other alloying elements improve its corrosion properties.

In pure zirconium the hydrogen evolved in the formation of the protective film diffuses into the metal surface and leads to hydride formation and breakdown of the oxide layer. The minor alloying elements in Zircaloy 2 prevent this and the hydrogen is evolved in molecular form and diffuses into the water. The bad effect of nitrogen is also decreased.

CONDENSERS AND HEAT EXCHANGERS

Apart from the wide range of waters that must be used for cooling and the consequent variability in corrosion characteristics, some of the most corrosive conditions are encountered in power stations and ships which draw their cooling water from estuaries and harbours. These waters are polluted to varying extents depending on location and the time of year. Thus water drawn from an area near the outfall of a sewerage works will be heavily polluted and will cause localised corrosion of condenser tubes. The attack started by polluted water, which is sometimes intergranular, may often continue even after changing to a clean water. The concentration of pollution, e.g. by hydrogen sulphide, which arises from the decomposition of organic matter, is likely to be greatest in late summer. The oxygen is largely used up by organic matter and sulphate-reducing bacteria become active.

Apart from hydrogen sulphide, which is probably the most powerful polluter of waters and can produce rapid perforation of materials such as brass, one of the most important materials is cystine. This is produced by the breakdown of organic material such as seaweed and may be present in both off-shore and in-shore waters, the concentration being particularly large in harbours and estuaries. It is a very efficient cathodic depolariser and can produce intense attack even in the absence of oxygen. If it is present in sufficient concentration, however, a protective film will form on copper alloys consisting of a copper-cystine complex. This film may break down or blister if it becomes too thick, e.g. under impingement, and gives the dangerous situation of a small anode and large cathode.

In the construction of condensers and heat exchangers the metals and alloys used are copper, brasses, bronzes and cupro-nickel alloys. The corrosion resistance of these latter materials depends both on the inherent nobility conferred on them by copper and nickel and the degree of protection afforded by the layer or film of corrosion product. The nature of this product, its continuity, adhesion to the metal surface and its ability to form, be maintained, and survive the action of erosive forces of high velocity sea water, determines the alloy chosen for a given use. The effects of velocity are complicated by the presence in water of entrained air bubbles and abrasive materials such as sand which can produce erosion and impingement attack leading to film breakdown and corrosion. The incidence of impingement attack on copper-alloy tubes is reduced when entrained air is removed from the cooling water. In other cases the presence in the water of dissolved and entrained air may actually be helpful in assisting the formation and maintenance of protective films on the more resistant materials.

The presence of iron corrosion products also helps certain alloys to form protective films and one case is on record where the introduction of ferrous sulphate into the cooling water at a power station reduced the attack on aluminium brass.

The use of pure deoxidised copper is limited to cases where purity of metal is essential. It has good resistance to practically all types of fresh water but it is not serviceable in salt water and waters polluted with sulphur compounds. (Above about 200°C there is a significant loss of strength). Impingement attack occurs when the velocity of the water is high and it does not form a protective scale, e.g. in soft water high in free carbon dioxide. Where this type of attack is expected, some means of reinforcement or protection are necessary or a change in material should be made. Arsenical copper containing 0.15 to 0.5 per cent arsenic has a higher strength and fatigue limit but still retains ductility. Its main use is in condensers and heat exchangers for use in fresh water. It is not recommended for waters containing hydrogen sulphide and other sulphur compounds, acid mine water and salt or brackish waters.

Copper is relatively sensitive to metal ion concentration cells and so the velocity of liquid across the surface is an important factor in increasing corrosion. Parts of a copper surface across which liquid is moving with a higher velocity become anodes and not cathodes as with a readily passivated metal such as iron. The resistance of copper to corrosion by rapidly flowing sea water is unsatisfactory and it is not suitable for condenser tubes in ocean-going ships or in power stations using tidal waters.

The use of muntz metal, 60/40 copper-zinc, is mainly confined to steam condensers operating at low temperatures and using fresh waters from rivers, lakes and wells. However, it has good resistance to hydrogen sulphide and other sulphur compounds and, with additions of lead, is commonly used in heat-exchangers cooled with sea water.

Red brass is also serviceable in heat-exchangers using fresh waters and is not susceptible to stress corrosion cracking and dezincification but is likely to be severely corroded in the presence of sulphur compounds.

The addition to brass of small amounts of alloying elements is beneficial and naval brass, 60/39/1 copper-zinc-tin, and Admiralty brass, 70/29/1 copper-zinc-tin, are used to resist sea water in exchangers providing the flow speed is less than 3 ft/sec. The addition of about 0.04 per cent arsenic to Admiralty brass inhibits dezincification but it is still susceptible to impingement. Nevertheless, with additions of arsenic, antimony or phosphorus, Admiralty brass is suitable for brackish or salt water at flow velocities up to 5 to 6 ft/sec.

In more severe conditions aluminium brass, aluminium bronze or cupronickels are used.

In aluminium brass, 76/22/2 copper-zinc-aluminium, the addition of aluminium assists in the formation of protective layers resistant to mechanical destruction, e.g. abrasion, and such brasses have a high resistance to corrosion-erosion. Aluminium brass with arsenic may be used in clean or polluted brackish and salt water up to 7 or 8 ft/sec.

Aluminium bronze, 95/5 copper-aluminium is used up to a similar velocity in polluted sea water.

Tin bronze is often recommended although its behaviour is not always as good as could be desired. It withstands fairly well scouring by abrasives, e.g. sand.

Copper-nickel tubes behave better than aluminium brass in polluted water. The addition of iron to the cupro-nickel assists in the formation of better protective films which are largely responsible for the excellent behaviour of the resulting alloys in sea water. The resistance is highest with the higher nickel and iron contents. In the use of cupro-nickels in clean and polluted sea water, the 10 per cent alloy may be used up to 8-10 ft/sec. but is not resistant to waters with a high sulphide content. Where long service life under severe conditions is required the 30 per cent nickel alloy should be used and has a velocity limit of 10 to 12 ft/sec.

In clean sea water moving at 11.7 ft/sec the rating of the copper alloys in decreasing tendency to impingement pitting is Admiralty brass, aluminium bronze, phosphor bronze with 8 per cent tin, followed by the cupronickels. The permissible rate of flow of sea water in heat exchange apparatus up to the appearance of strong corrosion erosion of the various copper alloys is: copper, 3.0, Admiralty brass, 6.0, aluminium brass, 8.0, and cupro-nickel, 26.0 ft/sec.

Nickel-bearing aluminium bronzes with additions of iron and manganese are used for ships' screws and are many times more stable towards erosion and cavitation than manganese brasses.

CONDENSATE

The corrosion of copper-zinc alloys by adequately treated steam condensate is less than 3.0 mdd. There is an appreciable increase in corrosion rates in untreated condensates containing oxygen, carbon dioxide and, sometimes, ammonia. A concentration of 30 ppm ammonia in steam condensate leads to deep grooving of unstressed brass and stress corrosion cracking of stressed brass.

70/30 copper-nickel alloys are used in boiler water preheaters handling steam condensate and boiler feed water. Their corrosion rate is less than 19 mdd but if the carbon dioxide content is high the rate is likely to be in the range 63 to 108 mdd. Steam condensate is corrosive to both nickel and nickel-copper alloys if the ratio of carbon dioxide to air in the dissolved gas is within certain limits.

Certain nickel-chromium alloys are highly resistant to hot boiler waters and steam condensate containing carbon dioxide and air. As an example, the corrosion rate of an alloy containing 13 per cent copper and 7 per cent iron is not more than 1.0 mdd in water at 70 to 135°C moving at a velocity of 8 to 15 ft/min.

The addition of copper to steel is beneficial in acid conditions and copper steels have been found to have a considerably longer life than mild steel when used with acid condensate.

Chapter IX

INHIBITION AND WATER TREATMENT

Before they are rendered suitable for public supply or for industrial processes many waters need some method of treatment, e.g. filtration,
chlorination, softening or distillation. Such treatments may increase or
decrease the corrosivity of the water for while softening or distillation
may render a hard water more corrosive, other treatments will reduce
the rate of attack on many metals. As an example, complex phosphates
which are used to prevent the deposition of carbonate scales are also
effective in reducing corrosion.

In many cases, additions to the water are made specifically to reduce or
prevent attack on a particular metal or combination of metals. Such
additives are referred to as corrosion inhibitors. If the metal remains
bright and to all appearances unchanged, behaving as a noble metal, the
inhibitor responsible for such a state is often referred to as a passivator.

The distinction between water treatment and inhibition is, however, a little
blurred. Thus sodium sulphite or hydrazine are not inhibitors although
they reduce corrosion by removing oxygen. Again, increasing the concentration of dissolved salts reduces the oxygen solubility and, in the absence
of movement, reduces corrosion rate. Yet a strong solution of sodium
chloride would not be considered as an inhibitor! In some cases the
terms 'water treatment' and 'inhibition' are treated as being synonymous.
The addition of silicate or complex phosphate is often referred to as
water treatment although such additives are also referred to as inhibitors.
Inhibition is best considered as a subdivision of water treatment on a par
with softening, deaeration or deoxygenation.

The degree of efficacy of inhibitors varies over a very wide range, from
the action of a passivator to the behaviour of calcium carbonate when the
attack may be appreciable before it is finally stifled. Such a material
would best be referred to as a restrainer, reserving the term inhibitor for
chemicals which permit very little or negligible attack. Many restrainers
are operative in lower concentration than inhibitors and are less likely to
affect the potability or toxicity of the water.

In preventing local attack or general corrosion, inhibitors and restrainers
serve a number of other desirable functions. They avoid the blockage of
pipes by corrosion products, keep the water clean for biological or
aesthetic reasons and maintain optimum conditions for heat transfer. The
latter characteristic may be just as important as the reduction in corrosion. The ability to maintain good pumping capacity in a pipe may mean
an appreciable saving in cost because the design diameter can be smaller
than is the case if allowance has to be made for roughness and reduction
in diameter caused by build-up of product.

Since iron is a common and inexpensive material of construction but generally has a poor corrosion resistance, most research on inhibition has been concerned with reducing the attack on iron. Many of the results are also applicable to other common structural metals.

An inhibitor may be considered as slowing down the anodic or cathodic reactions or, in many cases, both. It can do this in three ways, by slowing down the electrode process itself, e.g. dissolution of metal at the anode, by increasing the electrical resistance of the film present on the anodic or cathodic surface or by restricting or slowing down the diffusion of ions or molecules to the electrode.

CLASSIFICATION OF INHIBITORS

Inhibitors may be classified as <u>anodic</u>, <u>cathodic</u> or <u>mixed</u> according to whether they primarily affect the anodic, cathodic or both anodic and cathodic processes. The allocation of an inhibitor to any of these classes is only partially successful since most inhibitors affect both the anodic and cathodic processes. Nevertheless, such a division is useful in understanding the way in which inhibitors work.

Anodic inhibitors may be divided into oxidising inhibitors which can inhibit at concentrations of about 100 ppm and non-oxidising inhibitors which must be present at appreciably higher concentrations. The latter are all alkaline materials with some buffering capacity, i.e. a certain ability to maintain the pH value constant at an alkaline value.

The inhibition by oxidising inhibitors is not a direct function of oxidising power. Thus, chromates and nitrites act in the absence of oxygen, molybdates and tungstates are inhibitors but air must be present, and while pertechnetates* are good inhibitors at concentrations as low as 5 to 10 ppm permanganates have little inhibitive action.

Anodic inhibitors stifle the anodic reaction usually giving sparingly soluble substances as a direct anodic product. There is often no change in the appearance of the metal although it carries a very thin film which may be isolated by certain methods. In this class are included oxidising substances such as chromate and nitrite, which inhibit in the presence or absence of oxygen, and other salts such as hydroxides, silicates, borates, phosphates, carbonates and benzoates which are only effective in the presence of dissolved oxygen. The effective operation of such materials may be recognised by the more noble potential attained by the metal, which approaches that of the cathode. When present in insufficient amounts, such inhibitors, except benzoate, can be dangerous since they diminish the area of attack without decreasing appreciably the amount of metal dissolution and thereby produce intense local attack.

---

* Technetium, Tc, (element 43) was first characterised in the decay products resulting from the bombardment of molybdenum with neutrons or deuterons and is present in the fission products of uranium 235. The pertechnetate ion, TcO_4^-, is a very good inhibitor but very expensive.

Cathodic inhibitors stifle the cathodic reaction either by restricting the access of oxygen or by 'poisoning' spots favourable for cathodic hydrogen evolution. They include salts of magnesium, manganese, zinc and nickel which decrease the corrosion rate of iron and steel. The increase in alkalinity near the water-line by reduction of oxygen leads to the precipitation of the hydroxides of these metals as a reasonably adherent porous deposit which slows down oxygen diffusion. The presence of calcium bicarbonate in waters gives a general precipitate of calcium carbonate if the water is supersaturated or a local deposit on or near cathodic areas where the pH is high. The addition of lime to waters both raises the pH and serves as a cathodic inhibitor.

Cathodic inhibitors form a visible film on the metal, are not generally as efficient as anodic inhibitors and do not completely prevent attack. On the other hand they are less likely to intensify attack if added in insufficient amounts. Many waters contain both magnesium and calcium as natural constituents with inhibitive possibilities.

The products formed at the cathodes which, for the inhibitors mentioned, are colourless, can become coloured by interaction with the compounds formed at the anodes, e.g. rust. When these inhibitors are performing satisfactorily the potential of the metal may be as low or lower than the original corroding potential.

CHOICE AND USE OF INHIBITORS

The successful application of inhibitors requires adequate knowledge of their action and of the corrosion processes in the system under consideration. An inhibitor may prevent corrosion in one environment and increase it in another. The choice and concentration of inhibitor will depend on the type of system, the composition of the water, the temperature, the rate of movement, the presence of internal or applied stresses, the composition of the metal and the presence of dissimilar metals. The presence of loose scale, debris, crevices, etc., must also be taken into account.

The size of plant may vary from small closed installations using recirculated water to cooling systems using over a million gallons of water per day. On economic grounds alone the choice of inhibitor in the latter will be relatively restricted. The concentration of inhibitor will, in general, be greater the larger the concentration of aggressive salts. Thus the amount of chromate required will depend on the chloride and sulphate concentration which interfere with the formation of a passivating film on the metal.

In choosing an inhibitor, the corrosion engineer has to be familiar with the difference between 'safe' and 'dangerous' inhibitors. Safe inhibitors reduce the total corrosion without increasing the intensity on unprotected areas, while dangerous inhibitors produce increased rates of attack on unprotected areas (Fig. 37). Intensification of attack by the latter can occur in a number of ways: lack of sufficient inhibitor, the presence of enough chlorides and sulphates to prevent complete protection or of crevices and dead-ends into which renewal of inhibitor by diffusion is not rapid enough. Anodic inhibitors are, for the most part, dangerous inhibitors but one exception is sodium benzoate which causes <u>general</u> attack if full protection is not maintained. Cathodic inhibitors are generally safe but zinc sulphate is an exception and gives intensified attack along the water-line if its concentration is <u>too high.</u>

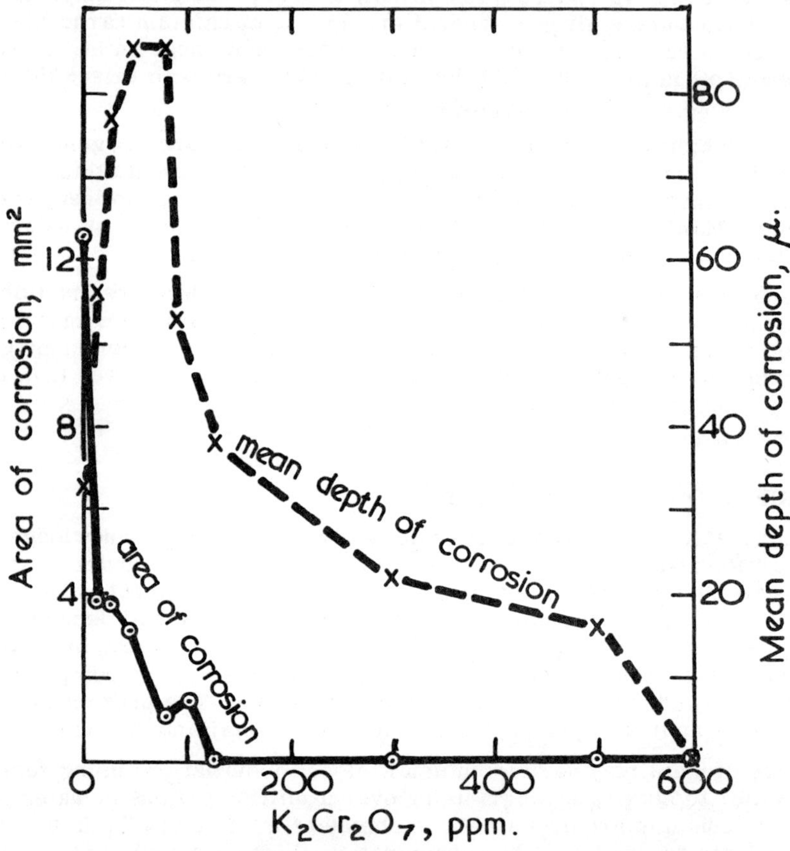

Fig. 37.   EFFECT OF POTASSIUM DICHROMATE ON THE CORROSION OF IRON IN AN AQUEOUS SOLUTION CONTAINING 30 PPM. SODIUM CHLORIDE AND 20 PPM. SODIUM SULPHATE.

(After Rozenfel'd, I.L. cf. Book by Putilova et al)

The increase in intensity of attack is readily understood if it is remembered that in most natural waters the total corrosion is controlled by the rate of reduction of oxygen on the cathodic areas. Reduction in the area of the anode can then intensify the attack and lead to perforation. When using chromate such behaviour can occur not only on iron and steel but also on zinc and aluminium.

The amount of inhibitor required will depend on the velocity of movement of the water and the relative ratio of solution volume to area of metal surface. In Chapter VIII it was explained that the higher the velocity of movement the thinner the diffusion boundary layer and the greater the amount of inhibitor reaching the surface. The overall effect is the same as increasing the concentration of inhibitor. In addition, water in movement is less likely to deposit any debris and screen the surface from the action of the inhibitor. Since inhibition is a dynamic process the inhibitor, at least initially, is being consumed in building up the cathodic or anodic film. Hence the safe inhibitor concentration is lower when the water volume is large than when it is small.

Great care is needed in applying the results of laboratory tests to industrial systems. Apart from the factors already mentioned an increase in temperature and the presence of dissimilar metals will both require an increase in inhibitor concentration.

The conditions and state of the metal surface and its accessibility to the inhibitor is also of the greatest importance. In industrial plant there are often places such as crevices or re-entrant corners at which the replenishment of inhibitor is slow. Although by good design these may be reduced to a minimum, some sites of this type may be unavoidable and should be considered when applying inhibition. There may be other places where foreign matter such as sand, metal filings, scraps of cotton waste used for cleaning, scale particles or rust, from another part of the plant, has settled. Such particles screen the surface locally from the inhibitor and may initiate localised attack, particularly with anodic inhibitors such as chromates, and corrosion can be very intense owing to the large cathode/anode area ratio. Such attack can be minimised either by cleaning the surface regularly or by increasing the concentration of inhibitor. On the other hand a number of inhibitors, mainly cathodic in action, will function satisfactorily on an already rusted surface and may operate with greater efficiency if a small amount of rusting has taken place.

A higher concentration of inhibitor is required if the surface is rough and has been sand-blasted or is covered with grooves and scratches. The same applies to surfaces under stress or on which impurities are present.

In deciding the amounts of inhibitor required to prevent corrosion by cooling water in industrial plant great care must be taken to take all these factors into account. Neglect of this creates the apparent discrepancies which exist between laboratory results and practical experience.

The compatibility of inhibitors with the medium or equipment must also be considered. The presence of appreciable organic matter may lead to the rapid depletion of oxidising inhibitors such as chromate, and bacteria may flourish in nitrite and phosphate solution. Again, soluble oils in anti-freeze mixtures can lead to rapid deterioration of rubber connection hoses.

The toxicity of the inhibitor and the question of its eventual disposal into rivers and public sewers must also be considered.

CALCIUM BICARBONATE

Calcium bicarbonate is the cheapest and most useful inhibitor for use in water supplies. Fortunately in many hard waters it is already present and it may be added to acid soft waters by a number of methods. Waters may be divided into four classes according to the carbon dioxide and carbonate content.

In the first class we have aggressive waters containing more carbon dioxide than is required to stabilise the bicarbonate. Such waters are capable of dissolving calcium carbonate, which may be present as a protective layer on the metal surface, or of preventing the formation of such a layer. Contrary to expectations, even water from a chalky source may contain aggressive carbon dioxide derived from decaying vegetation in the collecting area at certain periods of the year.

This dangerous carbon dioxide may be removed by passing the water through a bed of broken limestone or by cascading it to give a large liquid surface allowing the excess carbon dioxide to escape. The water is then closer to the equilibrium state with regard to solid calcium carbonate.

In the second class we have water which contains just enough carbon dioxide to stabilise the calcium bicarbonate present. Even though this may be the case, it is very rarely that the water is in equilibrium with the air since it contains more than the 0.5 ppm carbon dioxide contained in an equilibrated water. As carbon dioxide is lost to the air such waters will precipitate calcium carbonate. In terms of the Langelier Index such a water will have a positive value and if this is greater than about +0.5 the calcium deposited will reduce corrosion. A protective film may also form because cathodic alkali formed by oxygen reduction will displace the equilibrium towards the less soluble carbonate (Fig. 14). The smallest rise in pH will cause calcium carbonate to be precipitated and, together with the ferric compounds, will often form a protective scale.

In the third class, if the water contains less carbon dioxide than is required, the solution will be supersaturated in bicarbonate and will deposit calcium carbonate spontaneously on any solid surface. This may protect but can result in the production of a loose slime which far from preventing attack may actually intensify it by irregular deposition leading to the formation of differential concentration cells.

In the fourth class we have waters which are unsaturated in calcium bicarbonate and have a negative Langelier Index. Dissolution of any calcium carbonate will take place and if the Index is less than —0.5 the water may be relatively corrosive.

A method extensively used in the U.S.A. is to treat the water with sufficient alkali to render it just supersaturated in calcium carbonate and able to lay down a chalky coating. When the scale has reached the thickness of an egg shell the alkali is reduced until the water is just in equilibrium.

This method of treatment is unsuccessful if the water contains appreciable chloride and care should be taken in the application of the Langelier Index calculations to such a water. Occasionally the sodium content is greater than that of calcium which soon becomes exhausted at cathodic areas. The alkali then precipitates calcium carbonate at a distance from the cathode where it serves no useful purpose.

Soft waters of moorland origin often contain relatively little calcium and the acidity is due to both organic acids and carbon dioxide. Cascading will be clearly ineffective in this case and the water must be treated by passing through beds of broken limestone or, preferably, calcined dolomite. The dolomite is heated carefully at 500 to 600°C to decompose the magnesium carbonate to magnesium oxide but to leave the calcium carbonate unchanged. The de-acidification rate of dolomite is five to ten times that of marble because the magnesium oxide reacts rapidly with the free carbon dioxide to form alkaline magnesium bicarbonate and precipitate a thin film of calcium carbonate on the filter granules. This is soon dissolved by carbon dioxide to expose more magnesium oxide and the action continues. Both magnesium oxide and calcium carbonate play a part in the water treatment. Alternatively alkali may be added directly either as milk of lime, which is the cheapest, or as sodium hydroxide, which may be more convenient.

CHROMATES AND DICHROMATES

The film formed on iron and steel by chromate solutions approximates in composition to the passive film formed on stainless steel. The protective film, about 50 Å thick, consists of an insoluble iron compound such as iron chromate or a mixture of ferric and chromic oxides. The passivity of the surface may be recognised by the noble potential taken up by the metal. Although chromate initially inhibits the anodic reaction, in the course of time a continuous protective layer of ferric-chromic oxides is formed over the entire metal surface.

Both chromates and dichromates are widely used as inhibitors under service conditions and are very effective when properly applied. Owing to their toxicity, their use is restricted to industrial waters, but in this field they have been used longer and more extensively than any other inhibitor in their class.

The amount required depends on the variables already mentioned. It is advisable to use a relatively high concentration initially until a stabilised protective film has been formed. Once this has been obtained it is safe to reduce the chromate concentration but this should be done gradually, preferably spread over several months. The threshold below which the concentration should not be allowed to fall is about 120 ppm under favourable conditions. The chromate concentration can never be allowed to fall to zero without danger of rusting.

Dichromate should be used at a pH value of not less than 7.2 but chromate, which is more efficient, should be used at a slightly higher pH value of not less than 8.5. Since the dichromate is cheaper it is more economical to produce the chromate by mixing sodium dichromate and caustic soda in the proportion 100:27. To help in estimating the relative cost, 1 lb of hydrated sodium dichromate, $Na_2Cr_2O_7.2H_2O$, is equivalent in chromate concentration to 0.88 lb. of anhydrous sodium dichromate, $Na_2Cr_2O_7$; 1.09 lb. of sodium chromate, Na_2CrO_4; 0.99 lb. of potassium dichromate, $K_2Cr_2O_7$; 1.30 lb. of potassium chromate, K_2CrO_4, and 0.67 lb of chromic acid, CrO_3.

The chromate concentration required varies from about 100 to 10,000 ppm, the largest concentration being required in systems such as those containing bi-metal junctions, appreciable amounts of chloride, or those in which there is a danger of corrosion fatigue. In extreme cases, e.g. iron-brass couples in calcium chloride brine, complete protection may not be practicable. The concentration of chromate may be estimated by comparing the yellow colour of the treated water with that of solutions of known strengths. This is only possible if the inhibited solution does not contain any other colouring matter. In the initial period of treatment, tests should be made at least once a week.

The time taken to establish a stable protective film depends on the exposed area of the metal and, if the surface is large, it may take weeks or months. When the water is relatively high in chloride not only is a greater amount of inhibitor required but also the rate of pitting tends to increase if the chromate concentration drops below a certain minimum. This can be restrained by the addition of suitable amounts of phosphate as well.

A few examples from laboratory investigations and industrial practice will illustrate the range of concentration required under different conditions. In a slightly hard, stagnant supply water over 2,000 ppm chromate were required to prevent attack on abraded mild steel but in a hard water moving at 0.6 ft/min inhibition was obtained with less than 10 ppm chromate. Nearly 1000 times this latter figure was required to protect the complex cooling system of power rectifiers. Higher amounts are needed at elevated temperatures: a water that requires 500 ppm dichromate at 20°C may need as much as 2,000 ppm at 80 to 90°C and possibly 10,000 ppm if chlorides are present.

It is false economy to add insufficient amount of chromate since not only will chromate be rapidly destroyed by interaction with corrosion products but localisation of attack will result. If liberal additions are made at the start of the treatment there will be a saving in the long run since removal of chromate should cease after a few weeks.

For many practical purposes, e.g. the cooling systems of internal combustion engines, an initial concentration of 500 to 1,000 ppm is adequate and may be gradually reduced to 200 to 400 ppm. It is not possible to protect the metal entirely from attack in the region of the water-line but the effect may be reduced by increasing the inhibitor concentration to 5,000 ppm or by adding a suitable wetting agent.

In the use of chromate as an inhibitor the chief essentials are:

1. Sufficient concentration and the water made slightly alkaline.

2. As far as possible the metal surfaces should be clean and free from any debris so that there will be a uniform concentration of inhibitor in contact with the metal surface.

3. The water should be low in dissolved salts.

4. The water should be free from substances capable of reducing chromates to chromic salts which have no value as inhibitors. Thus chromates cannot be used economically if organic matter capable of being reduced is greater than a certain amount.

5. Chromate is toxic and also gives rise to skin irritation, <u>chrome itch</u>, with certain individuals. Hence exposure of the skin to these chemicals should be minimised and washing after handling is very important. It should be remembered that the discharge of appreciable amounts of chromate into sewers or rivers will not be allowed by Public Authorities.

NITRITES

Sodium nitrite is an effective inhibitor for iron and a number of other metals in a wide variety of waters ranging from fresh water, containing hardly any dissolved salts, up to brackish water and sea water. The amount needed increases with the chloride content and, although its use in waters of high salinity may be expensive, in such cases the choice of inhibitors is very limited.

The concentration needed to protect mild steel in distilled water is only 10 ppm but with a sodium chloride concentration of 500 ppm the inhibitor concentration should be raised to 300 ppm and with 3,000 ppm sodium chloride to 4,000 ppm. For 50 per cent sea water the concentration needed is 10 per cent.

Below pH 6 nitrite is relatively ineffective but its inhibitive properties improve with increase in pH value and reach a maximum in the range 9 to 10. Hence it can with advantage be used with borax which buffers the pH in the optimum range.

Like chromate it is an anodic inhibitor and passivator and acts by forming a stable film of ferric oxide which is continuously kept in a state of repair if adequate concentration of inhibitor is present. When the concentration is too low it is a dangerous inhibitor. In fresh waters enhanced attack takes place at the water-line if the nitrite concentration is less than about 1000 ppm. The tendency to this type of attack is greater at higher temperature. With insufficient nitrite the depth of corrosion is greater than in the complete absence of the inhibitor.

Unlike many inhibitors, sodium nitrite is effective even when the metal is already rusted, a very desirable property in industrial systems. In considering the application of nitrites to cooling systems it should not be forgotten that certain bacteria readily oxidise nitrite to nitrate which has no inhibitive action.

Besides steel, nitrite will also inhibit the corrosion of tin alloys, monel, aluminium and copper to a limited extent. The corrosion potential of tin

alloys is appreciably displaced towards more noble values in the presenc
of nitrite and this effect increases with concentration. In brine, 250 to
2,500 ppm nitrite suppresses the corrosion of tin but has practically no
effect on copper, which is in general only weakly passivated compared wit
steel or tin alloys.

Nitrite has practically no inhibiting action on the corrosion of brass since
the zinc reduces cupric ions to cuprous ions and so nullifies the oxidising
action of sodium nitrite. Furthermore the small amount of ammonia forn
ed in the initial reaction is considered to be responsible for the stress
corrosion cracking of brasses in the presence of nitrite.

BORAX

There is still some argument whether borax is an inhibitor or acts by vir
tue of its buffering capacity, which maintains the pH value in the range 8
to 9.5. It seems likely, however, that it is an anodic inhibitor similar to
sodium phosphate and carbonate and on iron forms a protective film of
which Fe_3O_4 and Fe_2O_3 are the main constituents. It is very cheap and
small amounts inhibit the corrosion of steel and zinc.

POLYPHOSPHATES

Polyphosphates and silicates behave in a similar way in the treatment of
waters: they both tend to buffer on the alkaline side of neutrality, and both
require the presence of oxygen. The use of condensed phosphates and
glassy silicates, separately or together, is the most economical method o
preventing corrosion in supply waters because of the small amounts required. Furthermore the treated waters are not harmful, and at the concentrations used the inhibitors have no influence on the taste or physiolog
cal effects of drinks or food.

Molecularly dehydrated phosphates, or polyphosphates, perform a dual
function in water treatment: they prevent scale formation and inhibit corrosion of steel and zinc. These condensed phosphates are a large group
of compounds containing chains or rings of phosphorus and oxygen atoms
joined alternately in the pattern:

$$-P-O-P-O-P-O-$$

These polyphosphates are complexing agents and with a divalent metal
such as calcium they form positively charged colloidal particles which
consists of numerous

$$-O-\overset{\overset{\textstyle O}{|}}{\underset{\underset{\textstyle O}{\|}}{P}}-O-\overset{\overset{\textstyle O}{|}}{\underset{\underset{\textstyle O}{\|}}{P}}-O-$$

chains joined by calcium ions or by the adsorption of calcium ions on
colloidal calcium phosphate.

wide variety of polyphosphates is in use, varying widely in solubility;
ie most common are pyrophosphate, $Na_4P_2O_7$, tripolyphosphate,
$a_5P_3O_{10}$, and hexametaphosphate. There are also a number of complex
lassy polyphosphates with a very low solubility. These glasses are
referably characterised by the molar ratio of sodium oxide to phosphorus
entoxide, i.e. Na_2O to P_2O_5. The chain length, n, is given by

$$Na_2O/P_2O_5 = \frac{n+2}{n}$$

or ratios 1.50 and 1.285 the chain lengths are 4 and 7 and correspond
» the tetra- and septa-polyphosphate respectively. A glass widely used
ommercially is sometimes described as 'hexametaphosphate'. If this
ere the case the molar ratio of sodium oxide to phosphorus pentoxide
ould be 1.33. It is usually about 1.1 corresponding to a chain length of
oout 20.

lthough the orthophosphate, Na_3PO_4, is an anodic inhibitor the action of
olyphosphates is predominantly cathodic. The positively charged col-
oidal particles migrate to the micro-cathodes where they build up a
lm which is self-retarding and which decreases the cathodic potential
nd restricts the supply of oxygen to the surface. There are also some
nodic effects since the film at the cathode contains iron as well as cal-
ium and phosphate and is built up more readily if traces of iron are pre-
ent in the water. The composition of the film is considered to be a
alcium-iron double phosphate with a solubility of less than 0.1 ppm at
H values in the range 5 to 9.

he formation of a thin protective film is not affected by normal varia-
ons in temperature or pH but its rate of formation is a function of the
ate of supply of phosphate to the metal surface. Thus circulation and
nixing of the water are important factors in determining the rate of
ccess of phosphate to the metal surface and it is very difficult to main-
ain the desired film in stagnant pockets.

or the prevention of scale deposits, 2 to 5 ppm are normally sufficient
ut the amounts required for corrosion protection vary from 3 to 150
pm (expressed as P_2O_5) according to conditions. As with many inhibi-
ors it is useful to start with an initial shock dose which may be decreas-
d once the film has been formed. Widely differing figures are quoted but
or use in cooling towers 15 to 37 ppm (as P_2O_5) are most common.

lthough polyphosphates alone will inhibit to a certain extent, the most
fficient effect is only obtained if an adequate concentration of divalent
netal ions such as calcium, magnesium or zinc is present. The required
mount is usually present in the water but with very soft waters lime must
e added. It is desirable to have a ratio of phosphate to calcium carbo-
ate not greater than 2 which corresponds to a phosphorus pentoxide/
alcium ratio of not greater than 3.35.

he presence of adequate amounts of divalent ions such as calcium or
inc also raises the limiting amount of chloride which may be present.
xcellent results have been obtained in preventing attack on mild steel

by solutions containing up to 1,500 ppm chloride ion by adding polyphosphate at a concentration of 50 ppm (as P_2O_5) and calcium or zinc at a concentration of 30 ppm. This method of inhibition is effective over a range of flow velocity from 0.04 to 7 ft per sec and at the latter speed the rate of attack at the high chloride concentration of 1,500 ppm was reduced from over 2,000 mdd to a negligible figure.

Dissolved iron is both a disadvantage and an advantage. On the one hand polyphosphates will be used up in forming a complex with the iron and indeed may be used to prevent discoloration from iron when added in the proportion of 2 ppm for 1 ppm of iron in solution. This addition must be made before the water is chlorinated or exposed to the air. On the other hand iron is needed for the formation of a protective film and helps it to be built up more rapidly.

In waters with a low calcium content use may be made of a glassy polymer of sodium-calcium polyphosphate which dissolves only slowly. On passing the water through a bed consisting of particles of suitable size, enough polyphosphate and calcium is taken up to render the water relatively non-corrosive to steel in hot and cold water systems. This treatment is effective in waters having a low silica and calcium content as well as in tuberculated systems. Other commercial mixed polyphosphate contain zinc and sometimes manganese metaphosphates.

Serious pitting can occur if the water is alkaline and underdosed. The danger of pitting is always less in acid solution possibly owing to the reductive dissolution of the oxide film. This will not occur, however, if mill scale is present. The inhibitor is most efficient at a pH below 6 and is best kept in the region 5.0 to 5.5 in an all-iron system but it should be raised to 6 or even 7 if copper and lead are also present in the system. Striking results are obtained in a mixed metal system as macroscopic cells facilitate electrophoresis of the complex ions.

The optimum dose for successful water treatment varies widely according to working conditions and this is a practical drawback to the use of condensed phosphates. A further disadvantage is that they hydrolyse to sodium di-hydrogen phosphate, particularly in hot solutions. When calcium ions are present this leads to the precipitation of calcium phosphate. If the temperature rise is appreciable with waters treated with polyphosphate, the formation of a calcium phosphate sludge or scale can promote serious corrosion. The reversion is less troublesome with zinc polyphosphate and when a protective film has already been formed.

In their industrial use in large systems there are two different methods of approach. If the water is not very corrosive the pH is adjusted by the addition of alkali to the verge of scaling and 2 to 5 ppm of polyphosphate are added to prevent troublesome deposition. This method has the disadvantage that deposition of calcium phosphate may take place on hot surfaces, although some workers report satisfactory protection over the temperature range 4-99°C. The second method is applicable to pitting waters, where it is an advantage to spread out the corrosion by depressing the pH, and has been used successfully in recirculating systems. The pH

lue is maintained at about 7, if necessary by the addition of acid, and a
gh initial shock dose of 100 ppm is used to lay down a protective film.
fter a time this may be dropped to 20 ppm.

very soft water with a pH of 6.7 protection of steel under conditions of
rbulent flow has been obtained in the cold by the addition of a mixture of
) ppm sodium polyphosphate and 25 ppm sodium-calcium polyphosphate.
romising results in protecting galvanised tanks from attack by very soft
ater have been reported using a mixture of zinc metaphosphate, which
ssolves quickly, and zinc-calcium metaphosphate which dissolves slowly.

he protective film formed by the polyphosphate treatment survives for
considerable period, up to 2 weeks, after the treatment has been stopped.
urthermore, tests have shown that this method of treatment may be
plied satisfactorily to already rusted surfaces, addition of inhibitor
ringing the attack to a stop almost instantaneously.

DIUM SILICATE

licates were first used about 1920 to combat the plumbosolvency of
rtain waters in order to prevent lead poisoning. Subsequently it was
und that in most waters small amounts of silicate retard but do not
mpletely inhibit the corrosion of iron, copper, brass, aluminium and
nc. The effect is usually more pronounced in hot rather than in cold
ater.

licates have the general formula $Na_2O.\, x\, SiO_2$ where x is the molar ratio
silica to alkali oxide. In waters with pH greater than 6 a value of x of
3 is recommended, but in waters more acid than pH 6 a value of 2 is
dvocated.

he mechanism of the inhibitive action is not clear but it appears to be
onnected with hydrolysis of the silicate to give negatively charged colloid
articles.* These may be silica, SiO_2, or more complex silicic ions of
e general formula $(m\, SiO_2.\, n\, H_2O.p SiO_3{}^{--})2P^{-}$, which have the pro-
rties of colloidal particles and migrate to the anodic areas where they
rm a film. The particles tend to coalesce into larger and larger aggre-
ates and finally form a protective gel over the whole surface. Although
is considered preferable to have the metal free from thick corrosion
roduct or scale it is also apparent that the protective effect of the gel
s considerably increased if it is mixed with products of corrosion. In
is case the protective film is produced by the mutual coagulation of
articles of the metal hydroxide and siliceous gel deposited on the metal
urface (Fig. 38).

he corrosion of hot water systems may be prevented by running the hot
ater through a receptacle containing glassy silicate or by adding a com-
ercial silicate liquor. In large buildings it is often both convenient and

---

Although many natural waters contain appreciable amounts of silica
is is generally in the crystalloidal form which has no inhibiting pro-
erties.

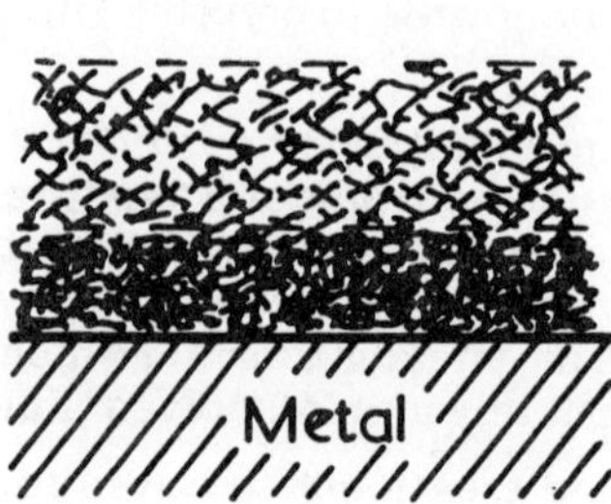

Fig. 38.   PROTECTION BY SILICATE

desirable to feed in the silicate near the cold water inlet by means of a proportional feed apparatus.

Corrosion is not completely inhibited and the degree of effectiveness varies with the water. The protection is usually satisfactory in pure soft waters and the effect is more marked in zeolite softened waters. Appreciable amounts of calcium and magnesium interfere with the production of a uniform silica film but this is avoided by the addition of about 2 ppm polyphosphate.

Calcium is believed to be beneficial and some American writers recommend addition of silicate up to the point where the water would be supersaturated in calcium silicate. Actual precipitation is avoided by the addition of polyphosphate. This is no doubt the reason why long-period tests with natural waters, which usually contain calcium, indicate much lower doses than laboratory tests which are carried out over shorter periods or tests when no calcium is present.

The amount required and the effect are not the same in all waters and the various dosages mentioned in the literature are doubtless due to differences in water quality and the way of assessing the attack. Initial dosage figures suggested vary from 8 to 16 ppm silica and when the film has formed throughout the system this amount may be reduced to 8 or even less. It is sometimes desirable to double the minimum dose every few months. If the chloride content is considerable it will be necessary to add 30 ppm at regular intervals.

For the protection of steel in aerated distilled water the addition of water glass with a molar ratio x of 2.3 was found to have a higher protective power than silicates with x values of 1.45 or 3.8. At a concentration of 700 ppm silica the corrosion rate was reduced by a factor of over 5 but in water containing 2,800 ppm silica, corrosion practically ceases.

Like all other alkaline inhibitors, when added in insufficient amounts, silicate localises and intensifies the attack in saline waters and it is probably unwise to use it for such waters without detailed investigation. On the other hand overdosing should be avoided since in concentrated

olutions colloidal particles are absent, but in dilute solutions an equili-
rium is set up between the ionic and colloidal forms. In the overtreat-
ment of water for domestic use excess alkali makes the taste unpleasant
and discolours some foods. 80 ppm silica in the water do not make the
water unpalatable.

Red water', caused either by oxidation of ferrous bicarbonate in natural
water or when iron and steel pipe is used in some soft waters, may be
corrected by the addition of silicate. Silicate treatment is effective in
hot water which causes most of the trouble from red water and corrosion.
It is also able to stifle corrosion under tubercles. With a relatively small
dose the corrosion of iron piping is reduced by 70 per cent.

Silicate treatment practically stops the dezincification of brass and the
corrosion of copper is retarded by regulating the dosage to give a pH of
about 8.

The addition of silicate is also recommended to water containing chloride,
sea water in particular, to prevent the corrosion of lead.

SODIUM BENZOATE AND ORGANIC INHIBITORS

Sodium benzoate is an efficient inhibitor of the corrosion of steel in
waters. Other metal benzoates, i.e. potassium, lithium, zinc and manganese also have inhibitive properties.

The benzoate is directly involved in the formation of a protective film
which is more loosely held on the surface than that produced by chromate.
Although benzoate is an anodic inhibitor it is not dangerous like chromate,
and if the protection fails corrosion spreads out over the metal surface evenly. Furthermore, benzoate does not promote water-line attack and, in fact,
protects the metal for some distance above the surface of the liquid
because of its tendency to 'creep'.

The main disadvantage of benzoate is the amount required and the pro-
hibitive cost for any system other than a small one such as the cooling
system of an engine. In a mixed metal system, a mixture of 1.5 per cent
sodium benzoate and 0.1 per cent sodium nitrite gives better results than
either inhibitor alone. Nitrite alone protects cast iron but increases
attack on soldered joints, but the combination of benzoate/nitrite protects
steel, cast iron and solder but not aluminium.

Monoethanolamine benzoate is one of the most effective corrosion inhibitors for steel in neutral aqueous solution but accelerates the corrosion
of copper, nickel and several other non-ferrous metals owing to the formation of complexes. Of the other similar organic salts, salicylates have
an inhibitive action but are not as good as benzoates which cover a wider
temperature and pH range.

The salts of cinnamic and nitrocinnamic acids also inhibit corrosion in
neutral solution and the salts of phenylacetic acid inhibit the corrosion of
iron in aqueous solution.

Triethanolamine phosphate, $(HOC_2H_4.NH_3)_3PO_4$, decreases the corrosion rate of steel as its concentration increases but increases the corrosion of copper. At a concentration of 1,400-2,800 ppm the corrosion rates of both steel and copper are negligible (Fig. 39).

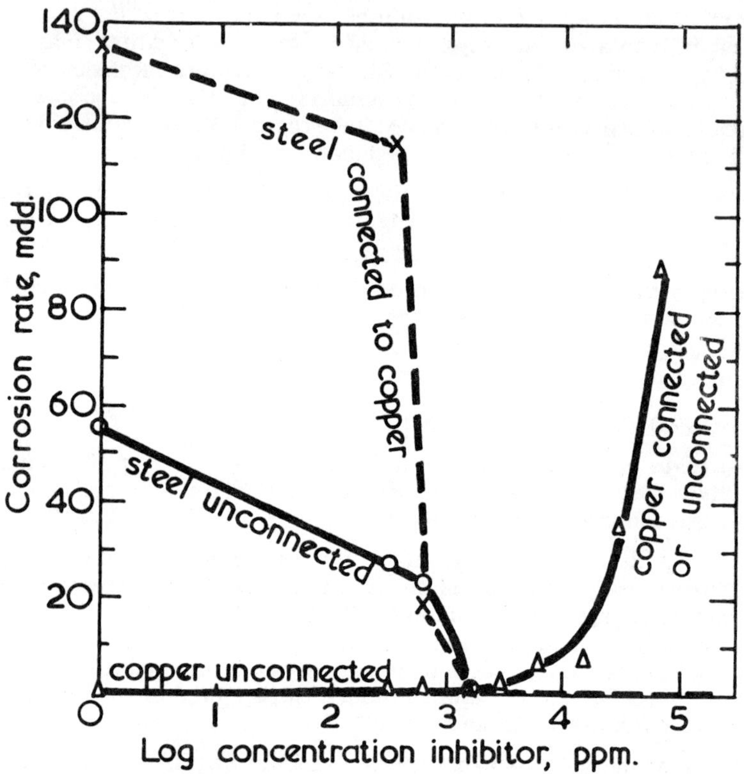

Fig. 39. CORROSION OF STEEL AND ELECTROLYTIC COPPER IN THE PRESENCE OF TRIETHANOL-AMINE PHOSPHATE.

(After Putilova, I.N., Balezin, S.A. and Barannik, V.P.)

Sodium mercapto-benzothiazole and benzotriazole are inhibitors for copper on which they form a protective coating. They also take care of copper ions in solution and prevent them plating out on more electro-negative metals such as aluminium.

Organic corrosion inhibitors have been little studied in aqueous environments. Their main advantage is that they have no range of concentration in which they accelerate attack or become 'dangerous'. Glucosates, dextrines, tannins and agar-agar, among others, have shown promise particuirly in water at high temperatures, 80 to 90°C, and where the cost is not rohibitive. Agar-agar at a concentration of 2,000 ppm in distilled water educes the corrosion of iron by 97 per cent and that of lead by 99 per ent. Egg albumin, dextrin and gelatine only reduce the corrosion of iron y a negligible amount.

igher alcohols reduce the general corrosion of steel in aerated sea ater and in potassium sulphate but give rise to pitting. The efficiency icreases with the number of hydroxyl groups in the sequence,

$$\text{sorbitol } (OH)_6 \ \rangle \ \text{erythritol } (OH)_4 \ \rangle \ \text{glycerol } (OH)_3 \ \rangle \ \text{glycol } (OH)_2$$

is significant that the ability of these alcohols to retard the oxidation of ydrated ferrous hydroxide varies in the same manner.

he tannins vary in behaviour: some combine readily with dissolved xygen, others produce films on metallic surfaces which may be protec- ve and others modify the form in which calcium carbonate is deposited nd give improved protection. Blending different tannins gives a mixture hich both reduces corrosion and scaling in the water-cooling systems of ngines, particularly diesels.

NHIBITOR MIXTURES

ften a mixture of inhibitors which function by different mechanisms perate better than either inhibitor used alone. Although, in general, one f the inhibitors is usually anodic and the other cathodic, this is not al- ays the case and both of the inhibitors may be anodic.

hosphate-chromate mixtures. Although both orthophosphates and chro- ates are anodic inhibitors, in combination they considerably reduce pitt- g at relatively low concentrations. The combined use of both phosphates nd chromates, at a concentration up to 100 ppm in solutions of low pH, is shown that over long periods both pitting and overall corrosion are reatly reduced. The retardation of corrosion is greater than that ob- ined by using either one of the inhibitors even at higher concentrations.

olyphosphate-chromate mixtures are better than orthophosphate-chro- ate mixtures and are best used at a pH of 6 to 6.5 to minimise rever- ion of the polyphosphate. The phosphate-chromate ratio is usually 2 to 1 nd the total concentration of inhibitor required is in the range 30 to) ppm. Thus on addition of 40 ppm polyphosphate and 20 ppm chromate e pits are shallower and smaller in number than in either 60 ppm chro- ate or polyphosphate when the pitting is quite deep. In an alternative rmulation used at a pH of 7.5 the chromate to polyphosphate ratio is 20 1, the polyphosphate being just enough to control scale formation.

he protective power is improved by the addition of zinc, say as zinc sul- ate, at a concentration of 1 to 2 ppm. This addition is particularly fective when the amount of phosphate is restricted by the calcium con- entration. If the zinc concentration is higher, problems due to precipita- on may occur. It is assumed that the cathodic film is rendered more impervious and more adherent by the presence of very insoluble zinc-phosphate compounds. Zinc, by restricting oxygen access in this way, also diminishes the possibility of crevice attack which is usually accelerated by single anodic inhibitors which are unable to gain access to the shielded surface.

BOILER WATER TREATMENT

The main objectives of feed water treatment are the prevention of scale formation and the prevention of corrosion. The scale is caused by the ingress of calcium and magnesium salts which deposit on the metal surface as calcium carbonate and magnesium hydroxide. Such scale is a poor conductor of heat hence heat transfer to the boiler water will be reduced and the temperature of the metal will increase. This may cause failure by reduction in mechanical strength or overheating under scabs of corrosion product.

To prevent corrosion, conditions should be created which give rise to the formation of a thin continuous layer of magnetite. This requires an initially clean metal surface, a low oxygen concentration, and a sufficiently alkaline boiler water. New or reconditioned boilers should first be cleaned of all oil, dirt, rust and scale in order to give the water treatment a chance to gain equal access to all parts of the metal surface. The boiler should first be <u>boiled out</u> with alkaline solution which emulsifies the oil and grease and loosens much dirt and foreign matter. Mill scale and rust are then removed by pickling with hot inhibited sulphuric, hydrochloric, formic, tartaric or citric acid. The clean surface is very reactive if left wet in the air and the acid is best expelled with nitrogen followed by conditioning with an alkaline mixture containing a deoxidiser, e.g. 0.1 per cent hydrazine.

In acid treatment care should be taken that no acid residues are left behind and hydrogen has not been introduced into the metal. Leakage of old boilers may occur on acid cleaning since some of the rust and scale may have been sealing leaks.

The maximum amount of oxygen that should be tolerated depends on the boiler pressure and may vary from 0.1 ppm in low pressure boilers to less than 0.007 ppm in high pressure boilers. The formation of magnetite is favoured if the water is alkaline, but unless special precautions are taken, there is a danger of caustic cracking or embrittlement by the concentration of caustic soda.

The raw water may require treatment to remove contaminants and dissolved salts and the removal of oil is very important. One of the main objects is to prevent the formation on the tubes of adherent scales which cause overheating and loss of thermal efficiency. The make-up requires softening, to reduce or eliminate the scale-forming salts of calcium and magnesium, or conditioning so that any material is deposited as a free flowing sludge and not a scale. Certain materials may be classified as scale formers or sludge formers. Calcium bicarbonate is a sludge for-

er but calcium sulphate is a scale former. Calcium phosphate and mag-
esium hydroxide generally form sludges but calcium carbonate and mag-
esium hydroxide can bake onto boiler tubes. Silicates are capable of for-
ing adherent scales which are very difficult to remove. Furthermore,
silica is volatile above 3000 lb per sq. in. and can form deposits which
damage turbine blades. Whether a sludge or a scale is formed depends
on the presence of nuclei on which the sludge can form.

Boiler condensate can be slightly acid or alkaline and may be contaminated with small amounts of copper and other metals but free from salts and
oxygen. It is important that fresh take up of oxygen should be avoided.

Water of high purity is required for high pressure boilers but many industrial plants are operating satisfactorily at pressures up to 1000 lb
per sq. in. with appreciable amounts of dissolved solids. In boilers working at 150 lb per sq. in. a hardness in the feed water of 40 ppm is acceptable but at 1500 lb per sq. in. even 1 ppm would not be tolerated. In general it is desirable to remove calcium and magnesium since they cause
scale formation and low heat transfer rates.

REMOVAL OF HARDNESS

Up to a pressure of about 200 lb per sq. in., water treatment is generally
confined to softening and if such water is only a small ratio of the total
feed, this treatment is adequate up to 400 lb per sq. in.

Precipitation processes. The removal of calcium and magnesium by the
addition of lime and soda is widely used, precipitating the calcium as carbonate and the magnesium as hydroxide and leaving an alkaline water.
Magnesium hydroxide tends to form positively charged particles which
may be flocculated by the addition of sodium aluminate. In a well-designed
plant, correctly operated, the hardness can be reduced to 10 ppm or less.
Some reduction in the total dissolved solids of most raw waters is obtained, much of the dissolved and combined carbon dioxide is removed and the
final water has an alkaline reaction. When operated at elevated temperatures and sometimes at a slightly positive steam pressure, the bicarbonates
are thermally decomposed and a large amount of the dissolved air is also
removed. This process can produce water acceptable as feed for certain
high-pressure boilers when no condensate is available.

Base-exchange. In base exchange methods the calcium and magnesium in
the raw water are replaced by sodium by passing through a bed of zeolite
which may be regenerated with common salt. The removal of the hardness
is more complete but there is no reduction in the corrosive tendency; this
may actually increase. When used with one of the precipitation processes
this method can give water of low hardness, low carbon dioxide and reduced in dissolved solids and alkalinity.

When heated in the boiler, sodium carbonate decomposes to give carbon
dioxide and caustic soda. The latter is an inhibitor but can also produce
caustic cracking.

PURIFICATION

In high pressure boilers it is desirable to work with a lower amount of
dissolved solids than can be achieved by softening. This can be done by
purification either by distillation or demineralisation, (ion exchange).

Distillation. A modern distillation plant will reduce the dissolved solids
0.1 ppm. and, with the latest vapour purification equipment, to 0.01 ppm.
Surface waters are usually first clarified and filtered but this may often
be omitted when the water is taken from the public supply. This method
purification is important when the water is very hard or impure or wher
the make-up is large, i.e. where steam is used for industrial processes
heating and no water is recovered.

Demineralisation. In power stations, where about 99 per cent of the wate
evaporated is recovered, the one per cent make-up is usually <u>deminerali</u>
ed or <u>deionised</u>, i.e. all ions other than hydrogen and hydroxide are remo
ed. Synthetic resins with acidic groups will replace sodium, calcium or
magnesium ions by hydrogen ions and other resins with alkaline groups
will replace sulphate and chloride ions by hydroxide ions. The resins,
when exhausted, are regenerated by acid or alkali respectively. Mixed b
units are often used to <u>polish</u> and improve the quality of the distillate fr
evaporators.

ALKALINITY AND pH

It is desirable to keep boiler water alkaline and the higher the pressure
the higher the pH value should be. The maintenance of a pH of 9.0 or
more reduces corrosion appreciably and encourages the formation of a
protective film of magnetite provided the oxygen concentration is sufficiently low. As the concentration of the hydroxyl ion is increased, the
solubility of oxides and hydroxides is reduced and leads to deposition
close to the metal surfaces. Conditions are favourable for the formation
of closely spaced nuclei of ferrous hydroxide or magnetite which form a
protective film. The transformation of the ferrous hydroxide, first form
to magnetite is more rapid if a catalyst such as copper or nickel is present.

One of the earliest methods of maintaining alkalinity and high pH was by
the use of caustic soda. This must be carefully controlled or it may lea
to disruption of the protective magnetite film and caustic cracking. The
latter is produced if the concentration of caustic soda rises to a value ir
the range 4 to 60 per cent, the amount of caustic required decreasing wi
increase in temperature and/or stress.

At boiler pressures below 200 lb per sq. in. sodium carbonate is often u
to control both scale and alkalinity. This is not the best practice since
high temperatures hydrolysis takes place to caustic soda and carbon
dioxide, the latter giving rise to acid condensate. Even at 200 lb per sq.
the rate of hydrolysis is so great that it is difficult to maintain enough
carbonate to precipitate calcium as a sludge and to prevent scale forma
tion.

ıe use of sodium phosphate, Na_3PO_4, which may be added to the feed
ıter or boiler water, increases the alkalinity and avoids the need to use
ıustic alkali. The main disadvantage in adding to the feed water is that,
ith hard water, a sludge of calcium phosphate deposits in the feed pipes.
 the boiler, phosphate is effective in the prevention of scale and if the
ıudge is sufficiently mobile it may be removed by regular blow-down. If
 is desired to reduce the alkalinity, disodium hydrogen phosphate,
ı2HPO4, which is the cheapest form of phosphate, may be used to replace
part or the whole of the trisodium phosphate, Na_3PO_4. If the pressure is
; high as 1700 lb per sq. in. other phosphates of lower solubility may be
·oduced.

 is often more convenient, e.g. when the water is hard, to add sodium
ılyphosphate together with caustic soda. If the temperature of the feed
ıter is less than 50°C no precipitation of sludge or scale takes place. In
 e higher temperature of the boiler the glassy phosphate hydrolyses to
·thophosphate and the calcium is precipitated as a phosphate sludge,
ydroxy-apatite). Combined treatment with sodium phosphate and caustic
ıda is also used to reduce the danger of caustic cracking.

ıe pH value of the cooled sample should give a definite pink colour with
 e indicator phenolphthalein, corresponding to a pH of at least 8. 4. The
kalinity to methyl orange should amount to 10 to 15 per cent of the total
ssolved solids.

 the many formulations, one commonly used at moderate pressures is
 e British Navy boiler compound containing 39 per cent sodium carbonate,
; per cent disodium hydrogen phosphate, and 13 per cent corn starch.

:her boiler compounds contain sodium carbonate, phosphate and some-
mes hydroxide, together with tannins, glucose or dextrin. The part played
· tannins is doubtful; some assist in the removal of the last traces of
cygen, some are considered to act as adsorption inhibitors and sometimes
rm a visible protective film, probably iron tannate, and others affect
ıcleation and favour the formation of a sludge rather than a scale.
sually a mixture or blend of tannins is required.

: pressures above 1000 lb per sq. in. greater consideration must be given
 the amount of dissolved solids in the feed and boiler water and, to mini-
ise the corrosion of copper and iron, oxygen, carbon dioxide and other
:id gases must be kept low. A more recent and improved method of treat-
ent is the addition of ammonia or amines which allow a closer control
ı pH value to be maintained. In pure water the amount of these chemicals
:quired to give a pH of 9. 0 is less than 0. 5 ppm ammonia, 2. 0 ppm of
·clohexylamine and 4. 0 ppm of morpholine. The amino-compounds have
 e advantage that owing to their volatility in steam, the condensate is kept
kaline and dangers of corrosion in the external circuits is reduced.

ınson boilers or once-through boilers working at temperatures above
 e critical point are the most difficult to protect because dissolved solids
 the boiler water must be extremely low.

ıme operators favour the use of pure water without any additives but this

could be dangerous since there is no margin allowed for the slightest
ingress of acidity or oxygen, both of which would interfere with the forma
tion of magnetite.

Ammonia is used to give an alkaline reaction and hydrazine to remove th
last traces of oxygen. Continuous and complete removal of oxygen is
absolutely essential otherwise troubles may arise from attack by ammon
on copper parts in the system. There is no take up of copper if the
ammonia concentration in the feed water is less than 5 ppm and oxygen i
kept below 0.02 ppm. In practice a still lower ammonia concentration is
adequate, e.g. in a plant giving steam at 610°C, 0.2 ppm ammonia is usual
Hydrazine tends to decompose to ammonia and this must be taken into
account when making the additions.

Cyclohexylamine and morpholine are important volatile sources of alkali
and may supersede ammonia since they possess specific surface active
and inhibiting properties. Hydrazine is both an oxygen scavenger and a
source of ammonia but these two properties are exclusive and cannot be
independently controlled by regulating the dose of hydrazine.

REMOVAL OF DISSOLVED GASES

Oxygen and carbon dioxide are the main gases dissolved in natural water
Small amounts of ammonia and hydrogen sulphide may also be present
derived from polluted cooling waters. Ammonia may also be present fro
the breakdown of hydrazine or added deliberately to the boiler water and
is corrosive to copper and its alloys in the presence of oxygen and carbo
dioxide. Hydrogen sulphide can arise from the breakdown of sodium sulphite.

The permissible levels of dissolved oxygen vary according to the temperature. In cold water 0.4 ppm is reasonable, at 70°C 0.1 ppm, at 200°C
(225 lb per sq. in.) 0.04 ppm and at 310°C (1430 lb per sq. in.) 0.007 ppm
The oxygen may be removed by mechanical or chemical means but the
latter is expensive and is generally only used to remove traces remainin
after mechanical deaeration

Mechanical methods. The oxygen, together with carbon dioxide, can be
removed <u>thermally</u> by spraying preheated water into an evacuated chamber. Alternatively the water is heated to near boiling point and introduce
into a deaeration chamber where it flows down a rack of trays against a
counter flow of air-free steam which carries off the liberated gas. In the
absence of any special deaeration plant, the oxygen content is reduced by
introducing the make-up at the condenser where low pressures exist. In
low pressure boilers it is usually sufficient to cascade the feed water int
the steam space of the boiler when the dissolved gases will flash off into
the steam.

Chemical methods. It is desirable in all systems to keep the oxygen concentration as low as possible. Where only a very low oxygen level can be
accepted use is made of oxygen scavengers to remove the final traces.
The choice lies mainly between sodium sulphite, Na_2SO_3, which has been

n general use about 30 years, and hydrazine, N_2H_4, which has been in use
or about a decade. Sodium sulphite reacts with oxygen with the formation
f sodium sulphate:

$$2Na_2SO_3 + O_2 = 2Na_2SO_4$$

he reaction is slow and may be inhibited by certain constituents present
n moorland water. The reaction is catalysed by very small amounts of
opper or cobalt which are often added intentionally although many boiler
aters will contain enough copper. Alternatively, use may be made of
ctivated carbon which operates by adsorption and concentration of the
xygen. An increase in temperature will also accelerate the reaction.

n order to remove all the oxygen there must be excess of sulphite present
ver and above the theoretical amount of about 8 parts by weight of sul-
hite required to remove one part by weight of oxygen. The recommended
xcess varies somewhat but typical values are in the range 20 to 40 ppm.
)ther workers use amounts of sulphite of 100 ppm and even greater.

 number of disadvantages are attendant on the use of this chemical. In
igh pressure boilers, particularly above 1000 lb per sq. in., it can decom-
ose to form sulphur dioxide or hydrogen sulphide which will give rise to
cid condensate. The presence of hydrogen sulphide in the steam can
ause rapid corrosion of nickel-bearing alloys. At about 250°C sulphite
may be reduced by iron or hydrogen, presumably to sulphide. Another dis-
dvantage is that the sulphate formed increases the dissolved solids in the
oiler water and more frequent blow-down will be necessary.

Iydrazine has the advantage that the products of its reaction with oxygen
re water and nitrogen and do not increase the dissolved salts:

$$N_2H_4 + O_2 \rightarrow N_2 + 2H_2O$$

his is, however, a very slow reaction and at temperatures below 140°C
ydrazine and oxygen can coexist in solution for quite a long time. Hydra-
ine is not normally added until after physical deaeration when the water
emperature will be at least 105°C.

he reaction rate increases appreciably with increased temperature and
he reaction times and hydrazine concentration are acceptable at about
00°C, i.e. well below the temperature at which the use of a scavenger
ould be justified. The reaction is catalysed by copper, silver and manga-
ese salts and also by passing the water through a catalyst bed of activa-
ed carbon.

heoretically hydrazine will remove its own weight of oxygen, which in
eaerated water will be much less than one ppm. To remove oxygen com-
letely an excess of hydrazine is required and figures ranging from 50 to
00 per cent are recommended. A further advantage of hydrazine is that
 may be used to control the alkalinity. Also, the cost is low and the re-
oval of oxygen is more efficient than by sodium sulphite, and much
maller amounts are required. Hydrazine is being used in boilers with

pressures ranging from 400 to 2, 500 lb per sq. in. It is easy to apply and can be easily controlled although oxygen is difficult to estimate in the presence of hydrazine.

The reaction is imperfectly understood but it appears to be a heterogeneous reaction occurring on the metal or oxide-covered surface. This can occur either by reaction between adsorbed hydrazine and oxygen molecules or indirectly, the ferric oxide being reduced by hydrazine to magnetite which then reacts with oxygen to reform ferric oxide.

Catalytic or thermal decomposition of hydrazine can give rise to ammoni according to the equation:

$$2N_2H_4 \rightarrow H_2 + N_2 + 2NH_3$$

or $\quad 3N_2H_4 \rightarrow N_2 + 4NH_3$

Metallic copper or magnetite accelerate the decomposition and copper ca also give rise to nitrous oxide, N_2O. If the residual hydrazine content is kept below 0. 2 ppm the ammonia in the steam will not exceed 0. 3 to 0. 5 ppm. In any case, in the absence of oxygen, the attack on any copper alloys will not occur and since hydrazine is added to remove oxygen the two should not be present simultaneously after passing through a high pressure boiler.

There is a toxicity problem and hydrazine should not be inhaled but the care necessary in handling is no greater than in dealing with concentrate acids or caustic alkalis.

CONTROL OF CAUSTIC CRACKING

The increase in concentration of caustic soda in various parts of the boiler, which are at high temperature and under stress, can give rise to caustic cracking or embrittlement. The concentration of alkali at which such reactions occur is dependent on temperature but the minimum caus tic soda is probably in the range 5 to 10 per cent. Such a concentration o alkali can be attained locally in leaky seams, e.g. of riveted boilers, or at the flanged joints of pipes, where the escape of steam causes concentratio of alkali. Such leakage in riveted seams is particularly likely if the rivet holes are badly aligned or the workmanship is faulty. In such conditions, stresses may also be present leading to stress-corrosion cracking. Concentration of alkaline boiler water can also occur in porous crusts or deposits, crevices formed by faulty tube manufacture, and crevices left after bad welding.

Caustic cracking requires high temperature, a high concentration of alka and tensional stress, particularly if they are non-uniform. To a certain extent the alkali concentration and the stress level are interchangeable. Thus at the ends of expanded tubes, where there is a residual stress, or where stresses arise due to injudicious caulking, and where the thermal stresses occur, the alkali concentration may be lower. This interchangeability of stress and alkali concentration means that there is no certaint of avoiding cracking even if sources of alkali are used which cannot concentrate to a dangerous level, e.g. sodium phosphate and ammonia.

After the alkali has reached a sufficiently high level it will attack the grain boundaries forming a number of branching cracks probably because, at the boundaries, the iron is subject to local stresses and is more anodic than the interiors of the grains. There is, however, special attack on the cementite present in the islands of pearlite. The main anodic product is soluble sodium ferroate with small amounts of the ferrite. Ultimately magnetite is thrown down, the alkali is regenerated and the attack continues. If the precipitation occurs at the tip of a crack the attack will be stifled but this will be a rare occurrence. The hydrogen formed in the first stage of the cathodic reaction is either transformed into molecular hydrogen or diffuses as atomic hydrogen into the metal where high pressures are built up in small cavities. The disrupting effect of this hydrogen pressure may play a part in the attack by strong alkali.

Sulphate-ratio. One method of controlling caustic cracking consists of maintaining the ratio of sodium sulphate to alkalinity above a certain value depending on the working pressure of the boiler. There is some doubt about the efficacy of this method but claims made on its behalf suggest that, as the boiler water becomes concentrated in the seams, sodium or calcium sulphate is precipitated and protects the highly stressed metal directly or plugs the seam.

Some users recommend that the ratio of chloride and sulphate to caustic soda should be five or greater. The maintenance of a high relative concentration of a neutral salt such as Na_2SO_4 is supposed to dilute the effect of the caustic by its presence in the concentrated boiler water.

Nitrate. The addition of nitrate is a widely accepted method of preventing caustic embrittlement. The ratio of the sodium nitrate concentration to the total alkalinity, expressed as caustic soda, is critical and should be 0.35 to 0.4. The nitrate may also be added as the potassium salt or as waste sulphite liquors which contain sodium nitrate.

One disadvantage is that hot nitrate solutions can cause intergranular cracking. However, nitrate cracking is more likely with nitrates of weak bases, e.g. calcium and ammonium, and is most pronounced in acid solutions so it presents little danger in the strongly alkaline boiler water. The effect may be compared to the behaviour of chromate which increases corrosion in acid solution but reduces corrosion in very slightly alkaline solution.

There is no generally accepted theory of protection by nitrate but it may provide an alternative cathodic reaction. If caustic cracking involves diffusion of hydrogen this may be prevented by the production of ammonia instead of atomic hydrogen.

Co-ordinated phosphate or zero caustic. This is a favoured method particularly in the U.S.A. Sodium orthophosphate, Na_3PO_4, is added to the boiler water until the ratio of sodium oxide to phosphorus pentoxide, i.e.

Na_2O/P_2O_5, is slightly less than 3, i.e. less than that which corresponds to sodium phosphate, Na_3PO_4. If the water is already alkaline, disodium hydrogen phosphate, Na_2HPO_4, is added instead. The ratio should not be allowed to fall too far below 3.0 or pitting instead of cracking may arise.

When evaporation takes place, solid sodium phosphate is deposited and the pH falls slightly. The pH cannot rise into the dangerous range, i.e. that corresponding to a saturated solution of sodium phosphate which may be just as alkaline as caustic soda. If the ratio is greater than 3, progressive evaporation leads to an increase in pH value and the situation can then become dangerous.

To avoid any deposition of solid, e.g. calcium phosphate, in the feed pipes, the phosphate may be introduced into the feed water as glassy phosphate which hydrolyses in the boiler to form the orthophosphate. It may be necessary to add caustic soda if the water is already acid or if insufficient alkali is present.

The main disadvantages of this method are its expense and the degree of control required, which is higher than that normally existing in association with a small boiler plant. Even in large systems under close control the variation in conditions from one part of the boiler to another may make the method unworkable.

Tannins (mainly quebracho extract) and butyric acid at a concentration of 0.5 per cent of the amount of alkali are also effective.

IDLE BOILERS

When out of commission boilers rust because of the unavoidable ingress of air. The method of preventing this depends on the length of time the boiler is idle. If the period is only a short one, say a few days, the boiler should be completely filled with an alkaline solution and an oxygen scavenger, e.g. caustic soda and sodium sulphite. The water with the added chemicals should be introduced at a low point in the system and allowed to overflow from the highest point which requires protection, e.g. the superheater. It is often difficult to avoid leaving air pockets owing to the geometry of the system. If these are too large, they will exhaust the oxygen scavenger and lead to attack at, or just below, the waterline. It is useless to add the chemicals after filling since they have little chance of reaching all parts of the boiler in adequate amounts. Sometimes chromate, e.g. 100 ppm at pH 8.5, is used instead of the sulphite, and hydrazine phosphate may also be used. The formation of ammonia by the latter may not always be desirable.

For long storage periods the boiler should be thoroughly dried and all valves closed. Drying may best be carried out by emptying the boiler while it is still under a few atmospheres pressure. If the boiler has already been emptied a small fire may be lit in the grate and drying accelerated by circulating dry heated air. Subsequently trays of a dehydrating agent, silica gel or quicklime, may be placed in the boiler to ensure complete dryness. Alternatively a volatile alkaline inhibitor, <u>vapour phase</u>

inhibitor, e. g. cyclohexylamine carbonate or nitrite, may be placed in the boiler when it is emptied.

CONDENSATE TREATMENT

Corrosion in condensers and return lines often occurs due to the presence of oxygen and carbon dioxide in the condensed water. Although the feed water to the boiler may be deoxygenated there may be leakage of oxygen into the circuit external to the boiler. Obvious methods to deal with this are by eliminating leakage or by the use of more resistant alloys. Where this is not possible or practicable sodium sulphite may be added to the condensate but a preferable treatment is to raise the pH value by the addition of volatile amines or the application of film-forming amines.

VOLATILE AMINES.

Neutralisation of the steam by adding a volatile alkali is found to be both effective and economic for reducing corrosion by condensate. Ammonia, one of the first used, may be added to the feedwater as ammonium hydroxide or ammonium sulphate, the ammonia being subsequently released in the the boiler. The major use of ammonia is where there is a low percentage of make-up and a low concentration of carbon dioxide. When the carbon dioxide is high, as frequently is the case in industrial plants, the level of ammonia required for neutralisation is high and the ammonia may be lost in the steam. This results in a drop in pH of the condensate and the re-appearance of attack of the ancillary circuit. Furthermore, ammonia is too volatile and requires close control to prevent attack on brass and copper. In relying on ammonia alone it is important to adjust the pH of the water to about 9.

Organic derivatives of ammonia, i.e. amines, have been found to have several advantages in the field of water treatment and are being used on an ever increasing scale. They are water-soluble and volatile in steam in which they circulate. The condensate formed is alkaline and all parts of the system are protected.

The two most important properties of such additives are their dissociation constant, which determines their effect on the pH of the water, and their volatility, which determines the relative amounts in the liquid and gaseous phases. Ammonia and cyclohexylamine favour relatively high concentrations in the vapour phase while morpholine favours the liquid phase in condensing conditions. Weight for weight, morpholine will produce a much smaller pH change than ammonia but since the volatility of morpholine is much less than that of ammonia, it may be used in spite of its greater cost.

All volatile amines are not equally efficient in preventing acid corrosion at the point of initial condensation. The vapour pressure of morpholine is such that even at low concentrations in boiler water it vaporises and the proper proportion condenses with the first drops of condensate which form in the system. Most other amines have vapour pressures which are too

low or too high and hence do not condense with the first droplets formed
and leave a critically important part of the system unprotected.

At 25°C carbonic acid is completely converted to morpholine carbonate at
a pH value of 7. 3 but at this figure the rate of reaction is slow and a
slightly higher pH value is desirable. Morpholine is more effective than
ammonia in the protection of turbines, is stable at high temperatures and
pressures and is evenly distributed. For effective corrosion control, a pH
of 8. 8 to 9. 0 is necessary (measured at room temperature). The amount
of morpholine or other additive required to maintain this depends on the
amount of carbon dioxide entering the system. Under favourable conditions the concentration which should be maintained is less than 5 ppm.
Morpholine is stable up to a boiler pressure of 2, 500 lb per sq. in. and to
a temperature of 650°C in superheated steam.

Cyclohexylamine and dicyclohexylamine may also be used for the prevention of the corrosion of iron by steam condensate containing oxygen and
carbon dioxide. For complete protection, the condensate must be maintained at a pH of 8. 2 to 9. 0. Cyclohexylamine has been used up to 570°C
and 2, 700 lb per sq. in. without any significant harmful degradation to
ammonia.

Several other amines have been tried but none of them are as effective as
morpholine or cyclohexylamine because they do not possess good neutralising ability, are insufficiently volatile or are unstable under boiler conditions.

The volatile amines may be introduced into the steam condensate system
by addition to the feed water, boiler, or return lines. Although there is
some advantage in adding the amine directly to the boiler or feed water it
has the disadvantage that an entire system is treated in order to protect
one localised section. Direct injection into the steam or condensate lines
is more economical. The amine losses are almost directly proportional to
the water and steam losses and any severely corroded section may be
given adequate treatment without wasting amine on the whole system.

FILMING AMINES

When the carbon dioxide content of the steam is high, or where the steam
is being used for some industrial process, the use of neutralising amines
is too costly and another group of amines must be used. These form a
barrier film on the metal surface and at the same time promote dropwise
condensation and improve the efficiency of heat transfer.

The adsorbed film on the metal is believed to have a thickness of only one
molecule with the polar or amine end of the molecules directly linked to
the metal by a strong bond. The degree of coverage or wetting of the
metal surface is a function of the strength of the metal-amine bond and
the orientation, shape and size of the molecule. Aliphatic amines with
straight chains of carbon atoms are more effective than those with branching of the non-polar part of the amine. The length of the carbon chain is
also important, amines with a chain length of 10 to 18 carbon atoms being
most efficient.

Octadecylamine, with 18 carbon atoms, is most frequently used, generally in the form of its acetate salt. Its wetting characteristics are improved by blending or emulsifying with a suitable wetting agent. Octadecylamine is not very volatile and is present in too small an amount to act by neutralising the carbon dioxide. The amine spreads over the entire metal surface and forms a protective and adherent film. This film is hydrophobic, causing water to condense as droplets, and acting as a protective barrier between the metal and corrosive condensate. Transfer of heat from steam to metal is better and insulation by corrosion product is kept to a minimum.

These film-forming inhibitors have strong surface-active or detergent properties and existing deposits and corrosion products are loosened. Since this action may lead to blockage it is advisable to clean out the system before starting treatment.

The recommended concentrations vary from 0.5 ppm to 25 ppm according to the formulation of the inhibitor but the average level of concentration is about 5 ppm. The formation of a protective film may take a significant length of time and, since these inhibitors are in the dangerous class, treatment should be started with a high concentration in order to lay down a protective film rapidly. Once the film has been formed the level may be reduced to that necessary to maintain the film in good repair after which interruption of the treatment for a few hours is permissible.

The material may be added directly to steam and condensate systems. The possibility of adding the inhibitors to the feed-water or to the boiler depends on the vaporisation characteristics of the amine used. The water temperature at the point of injection should, however, be at least 65°C in order to avoid premature precipitation, and the inhibitor should distil from the boiler to ensure protection at the points of condensation.

OTHER ADDITIONS

The addition of sodium silicate, polyphosphates or oils to condensate have had only partial success. Sodium silicate decreases the corrosive action of acid condensates by forming a protective film of silica on the metal surface or by neutralisation of the carbon dioxide by the alkali or both. Addition of oil in inadequate amounts may actually increase the corrosion rate instead of decreasing it. Polyphosphate treatment of return lines requires a rather long time to build up an impermeable protective film on the metal surface. To prevent dissolution of the film the dosage of polyphosphate must be maintained.

INDUSTRIAL COOLING SYSTEMS

The prevention of corrosion in cooling systems is of considerable importance not only from the point of reducing maintenance but also because adequate protection from rust formation will enable a high level of heat transfer efficiency to be maintained. Treatment is made difficult by the geometry of the system, the large capacity, dissimilar metals and by many variable factors: salt and oxygen content, flow velocity, temperature and volume of make-up. The problem of corrosion inhibition cannot be separated from that of algae and bacterial slimes which may actually be encouraged by some inhibitors such as phosphates.

Cooling systems may be classified as <u>closed</u> or <u>open</u>, according to whether the system is closed or open to the atmosphere, and, <u>recirculated</u> or <u>once-through</u>, depending on whether the water is re-used or run to waste. Recirculation is more common since once-through systems are only economic where the supply of water is cheap and inexhaustible as in ships or when a factory is close to a river or the sea. Closed systems can only be used where refrigeration is available or air cooling is possible as with radiators with forced air blast.

CLOSED COOLING SYSTEMS

These systems are less susceptible to attack by corrosion since the oxygen in the water is rapidly exhausted and the only cathodic processes involved are the evolution of hydrogen and activity of sulphate-reducing bacteria. In a completely closed system the hydrogen evolution type of attack is usually self-stifling owing to the formation of a layer of magnetite on any steel present. It may, however, be desirable to maintain a slight alkalinity, preferably by the use of sodium silicate or other inhibitor. To prevent the growth of sulphate-reducing bacteria a small amount of bactericide may be added, e.g. 5 ppm of cetyl trimethyl ammonium bromide or one ppm chlorhexidine. The use of chromate at a concentration of 1000 ppm in demineralised water at a pH of about 8.0 will perform both functions.

RECIRCULATION COOLING SYSTEMS

In many systems the water is recirculated and cooled by passing down towers or by spraying into ponds. The cooling obtained is mainly produced by the latent heat of evaporation. For maximum efficiency the largest possible area of water should be exposed to the air. The consequent evaporation not only increases the concentration of chloride and sulphate in the water but also ensures that the oxygen level is kept high. Furthermore, in industrial areas absorption of acid gases may cause the pH value to be as low as 4.0. All these factors increase the corrosivity of the water. Chlorination may increase corrosion if the concentration of chlorine rises above 0.4 ppm.

Concentration by evaporation may increase the original chloride and sulphate content by as much as 10 times. The hardness of the water will not increase in the same proportion since the excess will be precipitated as calcium carbonate. Make-up is necessary to replace loss due to evaporation, windage and water run off to keep the dissolved solids at some reasonable figure. The make-up may vary from hundreds to millions of gallons per day according to the size of the system.

The method of inhibition adopted is largely controlled by cost and possible dangers from toxicity. The latter is important in two respects; windage will carry the inhibited water over the neighbourhood while any bleed-off will affect sewers or rivers. Hence if any of the chromate formulations are used, the cooling towers will have to be shielded and care must be taken in the disposal of any water.

Corrosion may be retarded by the addition of lime to keep the water slightly alkaline or by the addition of silicates. Alternatively the water may be kept on the acid side of neutrality and polyphosphate additions made in combination with calcium or zinc if the amount of calcium or magnesium present is too small.

The condition of the metal surface will have an influence on the nature and particularly, the adhesion of any protective films formed. Treatment of new steel tubing for coolers and condensers is desirable. Any oil or grease should be removed by soaking in hot detergent, any rust and scale by sand blasting, and the tubing finally stored in inhibited water. Alternatively, after degreasing, pickling may replace sand-blasting.

In some areas cooling towers of aluminium alloy require no treatment but elsewhere the presence of heavy metals and high total dissolved solids leads to increased pitting and troubles due to scaling. Chromates at a concentration of 200 to 500 ppm and a pH value between 7.0 and 8.5 have been found satisfactory in preventing corrosion but the alloying constituents in the aluminium alloy may affect the continuity of the protective film. Polyphosphates are also effective with aluminium.

When both copper alloys and iron are present the problem of the corrosion of copper is usually small compared with that of the iron. It is important that the inhibitor used for the protection of the iron should have no adverse effect on the corrosion of copper. A mixture of chromate (organic is better than inorganic) and polyphosphate is more effective than either of the inhibitors alone or treatment with calcium carbonate. Both iron-copper and zinc-copper couples suffer the least attack in the presence of glassy phosphates which become adsorbed on the copper.

2-mercaptobenzothiazole (MBT) is a unique inhibitor in having a protective action on copper but not much effect on other metals. 1 to 2 ppm only are required in cooling towers and can be used with chromates or polyphosphates or both. The main disadvantage of this inhibitor is its high cost and its low solubility at low pH values. It may also interact with any chlorine present when the usefulness of each will be destroyed.

Chromeglucosates may be used in cooling tower systems particularly in very hot water, e.g. 80 to 90°C, where the oxygen content is lower and organic inhibitors have greater effect. The mechanism of action is not known but presumably it is a combination of the action of an organic inhibitor and a chromate. The disadvantages in the use of this type of inhibitor are the high concentrations required, the possibility of decomposition and also bacterial problems.

ONCE-THROUGH SYSTEMS

Systems in which the water is only used once are the most difficult to protect economically: invariably complete treatment is too expensive. They do have the advantage, however, that no concentration of aggressive salts occurs and the water used may be less than saturated in oxygen. In good cases, the water may be in a condition to lay down a thin protective layer of calcium carbonate. In other cases the occasional addition of lime, silicate or polyphosphate may perform the same function. Combined chromate-polyphosphate treatment can sometimes be economic.

Oil-in-water emulsions may interfere with the heat transfer because of the formation of an excessively thick oil film on the metal surface. They have, however, been used successfully in the cooling systems of engines. Their application to industrial cooling systems is of doubtful value since attack still occurs on areas where diffusion is restricted and in hard waters they form scums of calcium soaps. In addition they are less stable at high temperatures.

ENGINE COOLING SYSTEMS

The protection of the cooling systems of petrol and diesel engines is a most important field of corrosion prevention, as such engines are used in most forms of transport and for site power generation. The inhibition of the cooling systems is complicated by the number of different metals involved (Fig. 21), the variation in composition of the cooling water, the dangers of contamination and high oxygen concentration and the wide range of temperature and velocity to which the water is subjected. Up to nine or ten metals may be included in a cooling system and the corrosion of any one may be influenced by it being coupled to, or by the deposition of corrosion products from, another metal. Furthermore anything added to the water must have no damaging effect on the non-metals present, such as rubber, carbon and asbestos.

The composition of the water will vary in different districts and often will be saturated with oxygen. It is advisable to limit the chloride content and hardness and many U.S. railways have adopted a limit of 143 ppm hardness. Faulty joints may cause contamination either by exhaust gases, which create acidity, or oil which attacks hoses and seals, interferes with the action of inhibitors and forms sludges with scale deposits. The latter leads to local overheating and obstruction of the water passages.

The temperature of the water may vary from below ambient to 80°C or even higher if there is any malfunctioning of the circulation. The water may be stagnant or vary in velocity up to 5 to 10 ft/sec when flow may be extremely turbulent and cavitation may occur (Chap. VII).

Under winter conditions the ambient temperature may be as low as —45°C and antifreezes must be used. Ethylene glycol/water is the preferred mixture but additions of methyl or ethyl alcohol to water may be used although such mixtures have a low boiling point and consequent rapid loss. In operation, acids, mainly formic, may be formed by oxidation which is facilitated by excessive aeration, high temperatures brought about by the operating conditions or local hot spots and the presence of large amounts of copper or copper alloys. The concentration of antifreeze may vary from 10 to 60 per cent according to climatic conditions.

CHROMATES AND DICHROMATES

These are the most widely used inhibitors in diesel and gas engines and
are effective with a wide range of metals. The concentrations required
vary from 200 to 2, 500 ppm according to the water and operating conditions. Wherever possible, it is better to use a soft non-scale forming
water and with a hard water it may be an advantage to add a little polyphosphate as well.

A number of recipes for combined inhibitors have been suggested including chromate with nitrite or phosphate. The pH values recommended are
generally in the range 6. 5 to 9. 5, adjusted by adding soda ash or sodium
carbonate with the possible addition of borax to improve the buffering
capacity. When aluminium is in contact with copper the use of chromate
combined with sodium metasilicate at a slightly alkaline pH is effective.
When small areas of aluminium are in contact with large areas of other
metals it is desirable to buffer with borax. If the ratio of copper to aluminium is large no inhibitor will protect above 70°C.

The inhibitor most favoured in locomotive systems is probably sodium
chromate at a concentration of 2, 500 ppm (0. 4 oz/gal.) at a slightly alkaline pH. In view of the fact that chromate is toxic and dermatitic, some
operators have abandoned it in favour of other inhibitors.

SOLUBLE OILS

Oils in combination with an emulsifying agent, such as a long chain sulphonate, are effective inhibitors in reducing both general and galvanic corrosion. They are able to suppress the corrosion of iron and steel even in
the presence of appreciable amounts of chloride. The corrosion of aluminium is inhibited in waters containing up to 100 ppm chloride and attack
on aluminium-copper couples is also prevented.

The amounts recommended vary from 100 to 10, 000 ppm, the latter being
most favoured, and the ratio of oil to emulsifier varies from 10:1 to
100:1. When mixed with water they form a milky fluid which contains
negatively charged oil particles which migrate to anodic points where they
are deposited or are precipitated by interaction with iron ions. The oily
film formed over the whole surface not only acts as a diffusion barrier
but may also promote passivation by dissolved oxygen.

Their main disadvantages are that they belong to the dangerous class of
anodic inhibitors, can cause waterline attack and furthermore are difficult
to estimate by any simple spot checks. If the layer of oil becomes too
thick it can interfere with heat transfer. Soluble oils also have a tendency
to come out of solution and also to give troubles due to frothing. Probably
the greatest limitation to their use is caused by the fact that most oils
attack rubber and give rise to hose and gasket problems. This is particularly so with aromatic oils which in other respects are the easiest to use.

BORAX-NITRITE

A number of complex formulations containing from 5 to 10 constituents
have been used instead of chromates in diesel engines. The main inhibi-

tors in these mixtures are borax and sodium nitrite, the borax comprising
at least 50 per cent of the total concentration. Most inhibitors are not
generally applicable to all metal combinations and mixtures have to be
formulated with great care.

2-mercaptobenzothiazole or its sodium salt is often added for the protection of copper and its alloys. The pH values of the system, 8.5 to 9.5,
facilitate its solubility and enough of this relatively insoluble chemical
may be dissolved to provide adequate inhibition. Sodium silicate is added
for the protection of aluminium and it is fortuitous that borax gives a pH
of 9 to 9.5 in which the inhibition of aluminium is better than at neutral
values. Sodium nitrate is also added for improving the inhibition of solder
and aluminium and it also has some value in the inhibition of iron. Among
the other chemicals added are those to give protection against foaming
and scaling, to colour the treated water and to enable simple analytical
tests to be made.

Such formulations have the advantage of not possessing the toxicity problems of chromates, and are generally easy to apply but are relatively
more expensive.

In waters of medium and low hardness <u>tannins blended with soda ash</u> are
able to prevent both scaling and corrosion. Although they have shown
promise in diesel engines they are little used to-day.

INHIBITORS IN ANTI-FREEZE.

At low temperatures the cooling systems of engines must be protected
against freezing and for this purpose it is common practice to use an
ethylene glycol-water mixture. Inhibitors with a strong oxidising action
such as chromate cannot be used in such a medium because they react
with the organic coolant.

A mixture of triethanolamine phosphate (TEP) and sodium mercaptobenzo-
thiazole (NaMBT) will protect all the metals present in an engine cooling
system with the possible exception of nickel. TEP alone attacks copper
and forms a blue complex which deposits copper on steel and aluminium
surfaces and causes localised attack, but this is prevented by the presence
of NaMBT. Since corrosion products can remove TEP and NaMBT, it is
desirable to have a clean system before adding these inhibitors. The concentrations and the pH value of 6.9 to 7.3 must be maintained. (B.S. 3150.
1959).

Several variations of this formulation have been used, benzotriazole instead of NaMBT, cyclohexylamine instead of triethanolamine, and the addition of a nitrite to the mixture. Other variations are a combination of 0.5
per cent triethanolamine and 0.5 per cent alkali-metal nitrite, and a mixture of TEP and borate possibly with the addition of soluble oil.

Benzoate-nitrite is widely used with anti-freeze particularly in Great
Britain. In twenty per cent ethylene glycol a concentration of 1.5 per cent
sodium benzoate and 0.1 per cent sodium nitrite is recommended at a pH
value of 7.0 to 8.5 (Pls. 20 and 21). The nitrite is added to protect cast-iron.

(B.S.3151.1959). Since the inhibition is helped by heating it should be noted that if this formulation is used at and below ambient temperature the amount of benzoate and nitrite should be raised to 5.0 per cent and 0.3 per cent respectively. Sometimes the mixture does not protect aluminium but the attack is rarely dangerous.

In 50 per cent glycol, benzoate-nitrite is not as good as borax—NaMBT. If there is insufficient benzoate present there is some danger that the nitrite will increase the attack on solder or copper and its alloys in the system.

The addition of sodium tetraborate (borax) is the cheapest method of inhibiting glycol and acts as an alkaline buffering agent neutralising any acid formed in the oxidation of the glycol. The recommended concentration is 2.4 to 3.0 per cent and the pH value should be in the range 7.8 to 8.2. It is necessary to keep a check on the pH value to ensure that the buffering capacity is maintained. (B.S.3152.1959). It is advisable to have a small amount of sodium mercaptobenzthiazole (NaMBT) present to inhibit the corrosion of copper and prevent its deposition on other metals.

DOMESTIC WATER SYSTEMS

The treatment of domestic hot and cold water systems is limited by the effect of the inhibitor on the toxicity and palatability of the water and the impossibility of close control on dosage. Fortunately many waters have sufficient hardness for them to be relatively innocuous. More troubles are likely in the older type of system where the hot water supply is used both for heating and washing than in the more modern system where there is a closed primary circuit for heating and the hot water is obtained via a secondary circuit through a calorifier. Any closed primary system should be trouble-free since the oxygen is rapidly used up and the only trouble which may occur in a new system is an initial hydrogen evolution. This may be reduced by making the water alkaline, e.g. pH $\sim$ 8.5, by addition of caustic soda, lime or silicate. Such systems should be kept free from leaks so that the make-up is reduced to a minimum. When not in use they should not be drained. Troubles from sulphate-reducing bacteria can also occur.

Cost limits treatment to adjustments of the pH value and scale-forming properties. With some supplies, e.g. soft waters, it may be desirable to treat the hot water since the metal parts generally last less than one-third as long as parts in the cold water system. The temperature of hot waters should be restricted to a maximum of 60°C. Pitting becomes more likely at elevated temperatures since the calcium carbonate and other scales often break down locally, a process which is helped by the inevitable thermal cycling. If it is soft, the incoming water may be treated with silicate or polyphosphate or a combination of the two by means of a proportional feeder or the use of slowly soluble compounds.

The pitting of copper tanks and pipes is usually due to water of low pH with a high bicarbonate or carbon dioxide content. Simple adjustment of pH by addition of lime or silicate will usually prevent such troubles.

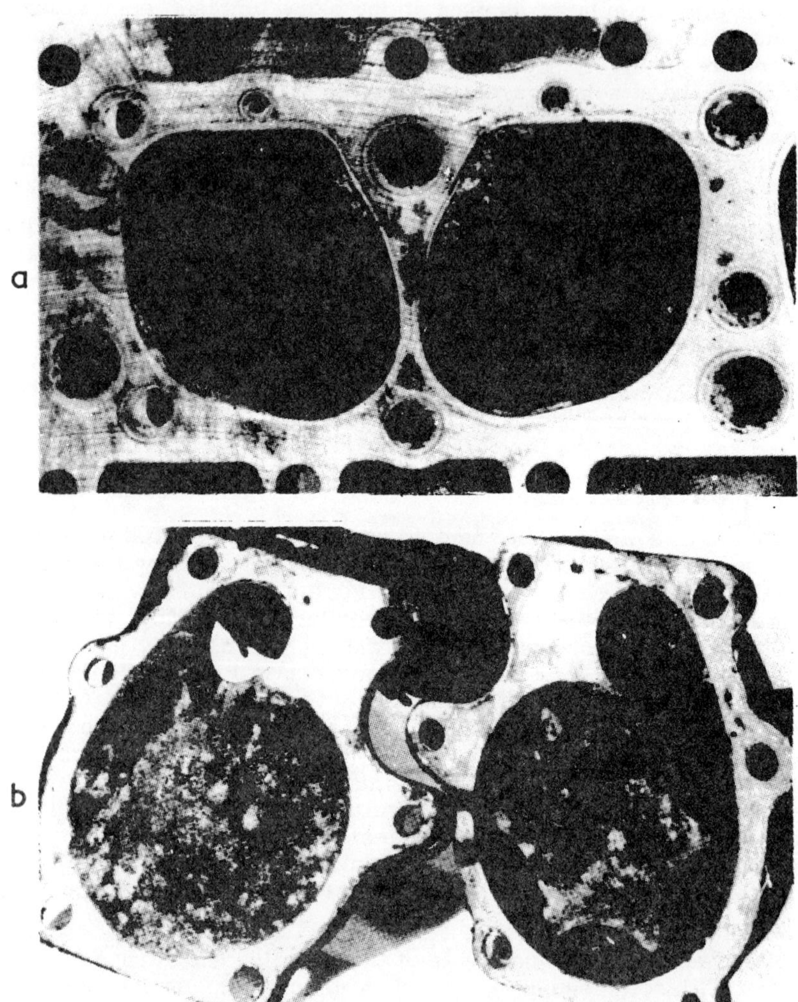

Plate 20. CYLINDER HEADS AND WATER PUMPS OF
VEHICLES USING ANTIFREEZE WITH AND
WITHOUT INHIBITOR.

a & b, 18,647 miles, 20 per cent ethylene glycol
without inhibitor.
c & d, 20,060 miles, 20 per cent ethylene glycol
+ 1. 5 per cent sodium benzoate + 0. 1 per cent
sodium nitrite.

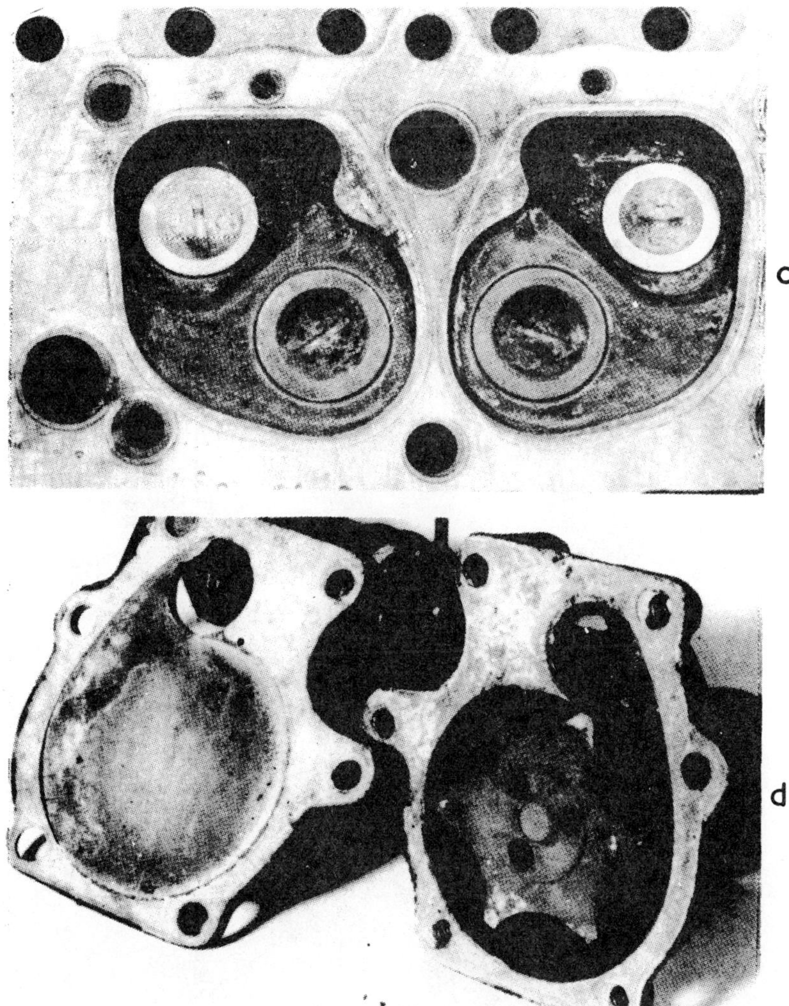
c
d

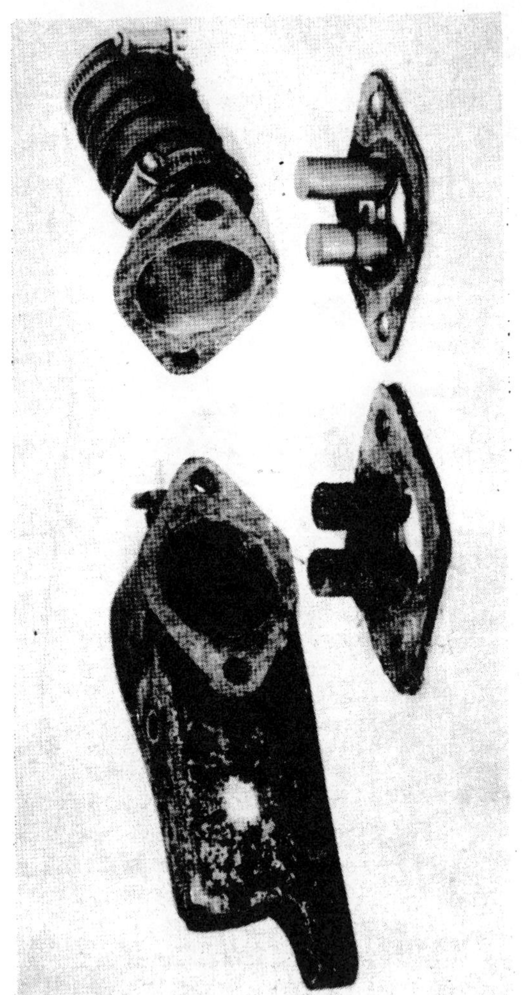

Plate 21.   SPECIMENS FROM THERMOSTAT HOUSINGS OF THE SAME VEHICLES AS SHOWN IN Plate 20.

Left: Two specimens of aluminium alloy separated by a copper washer.
Right: Cast iron specimen.

Chapter X

PROTECTIVE COATINGS

In many cases, metals are chosen on the grounds of cost or mechanical properties and their resistance to corrosion is poor or inadequate. Often conditions do not favour water treatment or inhibition and in such cases use is made of protective coatings which isolate the metal from the water or provide a surface more resistant to attack. The coatings may be metals applied by dipping, spraying or electro-deposition or non-metals, a class which includes the large number of paint formulations.

Apart from their corrosion resistance, such coatings should adhere well to the basic metal, be free from pores or cracks, have acceptable mechanical properties and, in many cases, present a smooth surface. In order to ensure good adhesion the primary requirement is that the surface should be free from grease, dirt and oxide. Cleanliness is just as important in the application of protective coatings as it is in conditions where the metal is exposed directly to water when oxides and other contaminants can produce local corrosion cells (Chap. I).

Many failures of platings and paint films are ascribable to inadequate pretreatment and it is a waste of money to specify a good paint when only a mediocre performance is assured by applying it to a poorly prepared surface. Any departure from the best practice results in an annoying loss of time and money.

PRETREATMENT OF METAL SURFACES

In order to prepare a clean metal surface it is necessary to remove not only grease and dirt but also oxide layers or manufacturing scales. This involves a combination of weathering or mechanical treatment, pickling and degreasing.

WEATHERING

The cheapest method of descaling steel is to leave it exposed to the weather. The combined action of wetting and drying and thermal fluctuation results in undermining followed by flaking of the mill-scale. This requires a long period of exposure which will vary with the weather conditions, the steel and the adhesion of the scale. In favourable cases a few months exposure will be adequate when most of the rust and remaining scale can be removed by wire-brushing. In the absence of weathering, mill-scale cannot be removed in this way. With low alloy steels the scale is likely to be more tenaciously held. In many cases, weathering can take two years during which time the steel is corroding at a rate which may be greater than 27.4 mdd (5 mpy) in an industrial district. Weather conditions on the day when exposure starts have an important effect on removing scale by under-rusting. Hence reliance on this method is open to serious objections.

MECHANICAL TREATMENT

Before application of paint or sprayed metal the main mechanical methods of treatment are sand-blasting, grit-blasting, flame-cleaning and wire-brushing and before plating it is usually necessary to grind and polish.

Sand and Grit-Blasting. In order to remove thick and hard coatings of scale, oxide or rust vigorous mechanical treatment is necessary. Sand and grit-blasting are widely used for this purpose and give a rough mat surface which serves as a key for the coating to be applied subsequently.

Sand and grit-blasting are effected by bombarding the surface with sharp-angled quartz sand, or iron or steel grit, introduced into a pressurised stream of air. The force can be regulated by the air pressure. The operator, particularly in the case of sand-blasting where there is a danger of silicosis, needs to be protected by an adequate extraction system and most blasting is carried out in suitably constructed booths. Portable equipment which can be used on site has been developed by surrounding the discharge nozzle with a suction tube to remove sand or grit immediately after impact.

It is common in many countries to use wet sand-blasting and the incorporation of inhibitors such as chromate or phosphate in the slurry has the advantage of protecting the cleaned surface from rusting.

Flame Cleaning. This method is inferior to abrasive cleaning in that it does not remove all the matter contaminating the surface, health hazards arise if the surface carries an old film of lead paint, and it is not easy to ensure that the surface has been adequately covered. The surface obtained in this way is particularly suitable for painting since the dry porous film remaining provides a good key for paint which is best applied to the still-warm surface immediately after cleaning.

Grinding. Grinding is carried out with sharp-angled materials such as silicon carbides or aluminium oxide which smooth and even out any rough surface. Grinding wheels are used or the grinding materials are carried on discs or bands or are coated on the inside of drums in which small articles may be tumbled. <u>Mechanical polishing</u> consists of grinding with finer media.

Wire-brushing. The work is here treated with rotating brushes of steel or brass wire (or with plastic bristles) while water or oil as a lubricant is slowly dipped on to the surface. Wire-brushing by hand is an important procedure in the renovation of paintwork when loose paint and corrosion products should, as far as possible, be removed.

DEGREASING

Degreasing is a most important part of the surface treatment and may be
carried out by physical or chemical methods. In many cases, e.g. in painting, degreasing by solvents is adequate, but for work in aqueous solution,
e.g. electroplating, the metal parts must be sufficiently clean to be wetted
by water and chemical or electrolytic methods are then necessary.

Solvent Degreasing. This may be carried out by the application of organic
solvents which dissolve organic and inorganic greases and, for the purpose
of safety, have a flash-point of not less than 38°C. The desirable properties of such solvents are a large dissolving power and degree of saturation,
a high solution rate, a high degree of penetration, the ability to neutralise
finger marks and negligible corrosive action on the metal surface. Petrol,
trichlorethylene and perchlorethylene are most commonly used. Adequate
precautions should be taken to prevent operators inhaling the vapour; it
is usual to carry out the work in closed tanks.

In large undertakings it is better and more efficient to degrease in the
solvent vapour since the latent heat of chlorinated hydrocarbons enables
large amounts of solvent free from grease to condense on the metal.
Even after this kind of degreasing, the wettability of articles is not adequate for electroplating.

Chemical Degreasing. In hot degreasing by strongly alkaline solutions,
organic fats and oils are broken down into fatty acids or glycerine which
are soluble in water. This method is not satisfactory for removing inorganic fats or oils used in modern workshop practice.

In degreasing, a number of processes occur and, in order that the full
value of such a method may be realised, various factors such as wetting
power, surface tension, emulsifying power and dispersive power have to
be considered. A modern degreasing solution will contain an alkaline
salt (soda, caustic soda, water-glass or trisodium phosphate), an emulsifying agent such as a fatty alcohol sulphonate, water softening agents such
as polyphosphate, and surface active organic compounds which affect the
surface tension of the solution.

Chemical degreasing is not always successful and in recent years ultrasonic methods have proved very satisfactory for small items. High frequency vibrations produce a very strong stirring action and both dirt and
grease are mechanically removed even from very complicated parts.

Electrolytic Degreasing. An article which is subsequently to be electroplated is degreased electrolytically by making it the cathode in an alkaline
solution. The cleaning is by combined chemical and mechanical action;
in the strongly alkaline zone at the cathode any organic grease present is
saponified and both dirt and grease are detached from the metal surface by
the strong hydrogen evolution. Care should be taken in applying this
method to metals which are susceptible to hydrogen embrittlement. Apart

from its efficiency, this technique has the advantage that it requires only a fraction of the time needed for dip greasing.

A more recent method is two-phase emulsion cleaning. Such materials as petroleum and oleic acid are converted into a water soluble emulsion by the addition of triethanolamine. When diluted 8 to 90 times by water this makes a very useful cleaning solution.

PICKLING

Chemical Pickling. In order to free metal from rust, oxides and scale, pickling in acid is often used. Attack on the bare metal is restricted by additions to the acid of inhibitors which are mostly organic compounds. To ensure that the attack should be even and efficient, it is necessary for the metal surface to be as free as possible from grease. Any of the common mineral acids may be used and, after pickling, efficient washing in hot water or rinsing in dilute alkali is essential in order to remove any traces of residual acid. Although more expensive, there is great advantage in using phosphoric acid for pickling such metals as aluminium and steel before painting; a phosphate film is formed on the surface which is not only a temporary protective but also forms a good bond for paint.

On materials to be electroplated, a bright-pickle is often preferred since in many cases this serves to give a bright electroplate. This should not be confused with chemical polishing which, like electropolishing, leads to a levelling of surface irregularities.

Molten salts are also used for pickling metals although a subsequent acid pickle is necessary since the mill-scale is often only loosened. The advantages of such a method are that there is no necessity for preliminary degreasing or mechanical removal of mill-scale, and there is no embrittlement by hydrogen and no attack on the metal. Cast iron may be treated cathodically and anodically at 450-510°C. The cathodic treatment removes the oxide, particularly the forging scale, by reduction, and the sand mechanically, while the anodic treatment removes carbon and other easily oxidised substances as well as occluded sand.

Yet another type of descaling is the treatment of parts in a sodium hydride solution.

Electrolytic Pickling. In practice, pickling is often carried out by making the metal cathodic or anodic in a suitable bath, both of which cause the mill-scale to loosen and crack off. An important advantage is the reduction in the time of treatment compared with chemical pickling and, although there is vigorous hydrogen evolution in cathodic pickling, hydrogen embrittlement is much lower than in chemical pickling. When it is desirable to avoid any danger of embrittlement anodic pickling is preferred.

It is important to note that during chemical pickling, working in confined spaces without adequate exhaust should be avoided. Hydrofluoric acid can produce severe damage to the skin and eyes and in such work protective glasses and rubber gloves should be worn.

METAL COATINGS

If a metal has an inadequate resistance to corrosion it is not always necessary to change to a more resistant, and usually more costly, metal. Often the basic metal, which may be chosen for its mechanical properties, can be coated with a second metal which is chosen for its corrosion resistance. Such coatings may be applied in a variety of ways; by dipping, electroplating, spraying, diffusion or cementation, or by mechanical cladding.

SURFACE PREPARATION BEFORE PLATING

Even after the surface of the metal to be coated has been cleaned by the procedures already outlined, it may not be possible to get an adequate deposit due to the physical state of the surface. The pickling of ferrous metals results in an enrichment of the surface in carbon. On metals such as aluminium an oxide layer is present. Several measures may be taken to remove these films, e.g. electroplating at very high current density or by the use of a strike bath with a high covering and throwing power. The latter has a low current efficiency but the plating period is only a few minutes.

Plating on magnesium, zinc and aluminium is difficult since they are either too active, or they are passive. In both cases adhesion of the deposit is bad and a suitable pretreatment or a modified bath composition is necessary.

On aluminium the main purpose is partially to dissolve the oxide layer and to create a network which acts as a key for the coating metal. The surface can be suitably treated in a zincate etch bath when aluminium is dissolved and zinc is deposited. The same object may be achieved by dipping in a solution of cyanide or soda or in an alkaline bath containing brass or copper which deposit on the aluminium. After treatment in phosphoric acid aluminium may be plated directly with nickel although a copper deposit is generally applied before plating with another metal.

Magnesium is degreased in a chromic acid solution containing fluoride and the surface is either activated by dipping in acid solution or coated with zinc by dipping in a weakly alkaline solution containing zinc sulphate and fluoride. The layer of zinc acts as a key for the subsequent electroplate, the deposition of which must immediately follow the pretreatment.

Zinc and its alloys may be directly nickel plated in a special bath with a pH between 5 and 6 and containing a large amount of a conducting salt such as sodium sulphate or an organic salt. It is preferable to give a good copper coating by dipping in an alkaline copper bath, to buff this deposit well so that it is compacted and then to nickel plate in the ordinary way.

For plated aluminium to be corrosion resistant the coating must be very compact and free from pores since the potential between the basis metal and the coating may be very large. This danger is even greater in the case of magnesium or zinc. Due consideration should always be given to

the relative potentials of the coating and basis metal in order to avoid
increasing the attack on the metal to be protected.

METAL DIPPING

Metals may be coated with another metal by dipping in an aqueous solution
of the second metal or its melt. Deposits obtained from aqueous solution
are generally very thin and do not confer any protection against corrosion.

More recently, 'current-less' methods of dip coating have been developed
which give thicker deposits. The solutions used contain metal salts, a
reducing agent and a catalyst. In the case of nickel plating such a bath
contains a nickel salt, hypophosphite and nickel or steel as a catalyst.

In general, however, dip coating is carried out with molten metals. This
method, the oldest method of protecting iron, is restricted to coating with
metals of low melting point such as tin, lead, aluminium and zinc. The
molten metal is very reactive and this leads to the formation of an alloy
layer between the parent and deposited metal which is often undesirable
so that steps must be taken to restrict its formation. Thus zinc and
aluminium give alloys which enhance adhesion but make the coating very
hard and brittle and the treated sheet cannot be worked since the coating
easily parts from the metal on bending. It is essential the zinc coating is
free from "dross" which will readily evolve hydrogen on contact with
water and so create an explosion hazard in closed vessels.

The nature of the basis metal is important and, on iron, both graphite and
sulphur have an influence on the diffusion and adhesion of the coating.
This may be overcome either by decarburising the surface, making it
sulphur-free or by depositing electrolytically an intermediate layer of
iron.

The control of impurities in the molten metal is very important. In the
case of zinc it is favourable to have a lead content of 0.75 to 1.25 per
cent and a cadmium content of 0.1 to 0.3 per cent but unfavourable to have
iron present at any concentration. The addition of aluminium, tin and
antimony influence the crystal structure.

Lead does not alloy with iron and in order to coat steel it is necessary to
add metals such as tin and antimony which readily alloy with both lead
and iron. The common coating, known as Terne plate, contains from 15 to
25 per cent tin and possibly additions of antimony, silver or zinc.

ELECTROPLATING

Electroplating has advantages over dip coating in the range of metals that
can be deposited, the saving in metal, the ability to produce thinner coatings and the absence of any brittle intermediate layer. Of the thirty
metals that can be electrodeposited only about fifteen are of technical
importance and many of these are mainly used for decorative purposes or

for protection against atmospheric corrosion. Plated metals are not used on any large scale in treated and natural waters.

The desirable properties in an electroplate are good adhesion, freedom from porosity, a smooth surface and good mechanical properties. These are only attained by use of the right bath and good plating procedure. The bath composition is very complex: besides soluble salts of the metal to be deposited, which may be simple salts of an organic acid or complex salts such as cyanide, it contains salts to lower the electrical resistance of the bath, an anodic depolariser and smoothing media. Wetting agents are added to prevent hydrogen or suspended particles from settling on the cathode and so cause porosity. Agents are also added to improve the throwing power of the bath in order to give a more even thickness of coating on irregular objects. The current density must be within the range specified for the particular bath, the bath must be stirred, often by oscillating the rack on which the objects being plated are mounted, and the electrolyte should be continuously filtered.

An increase in temperature enables higher current densities to be used and leads to deposits which have a coarser crystallinity and a softer texture. The plating bath is usually operated at the highest temperature possible. In order to keep the metal concentration constant, soluble anodes are used and since small amounts of impurity can have an adverse influence on the plating, the metal used for the anodes must be of the highest purity. To prevent anodic passivation depolarisers are added, e.g. chloride to a nickel-plating bath.

An important modern development is brush plating which enables large areas of metal to be plated in situ. An extremely high current density from 100 to 500 amps per sq. dm. is used and plating is carried out by moving an anode, which is a tampon or brush and carries the electrolyte, over the metal to be plated.

Copper. Copper coatings are rarely used as such but they play an important role as an intermediate layer since they are soft, and polishing or buffing seals all the pores in the basis metal. From the corrosion point of view, however, a copper coating is undesirable since it is almost invariably more noble than the parent metal.

Copper is frequently used between the basis metal iron and a nickel coating. If the coating is damaged sufficiently to expose the iron the severity of attack is greater than if the iron were nickel plated directly.

Brass. Brass is preferable to copper as an intermediate layer since copper in the brass diffuses into the overlying nickel and leaves the intermediate layer enriched in zinc, which is an advantage from a corrosion point of view. A coating of brass is often put on steel that is to be rubberised since the adhesion of rubber is thereby considerably improved.

Cobalt. Although it is expensive, cobalt has advantages over nickel in being harder, having a higher resistance to friction and wear and a very high corrosion stability.

Nickel. Nickel is one of the most important metals used in plating since it is hard and even relatively thin layers give good protection. Mat nickel coatings on steel always behave better than bright nickel which often has nickel sulphide, produced by decomposition of the wetting or brightening agent, occluded in the plating.

Electrodeless nickel plating may be produced by dipping in a hot solution of nickel chloride containing sodium hypophosphite as a reducing agent. A very even coating over parts of complicated profile can be obtained at a deposition rate of 0.25 mil per hour. The fine-grained deposit contains 6 to 11 per cent phosphorus and has a laminated structure. On heating above 400°C a mixed phase of crystalline nickel and a nickel phosphide, Ni_3P, is obtained. In 3 per cent sodium chloride the corrosion rate is 0.03 mdd.

Nickel Alloys. Nickel-zinc alloys cannot be deposited directly but they may be deposited separately and then alloyed by heating for 6 hours in the range 500 to 750°C. A tin-nickel alloy containing 65 per cent tin is very corrosion resistant and may be prepared in a similar way.

Zinc. Zinc is the most important metal in the protection of iron since not only does it have a better corrosion resistance but when damaged is able to protect cathodically the exposed basis metal. It is a debatable question whether hot dipping or electroplating is the better method but the latter has the advantage of not giving any intermediate alloy layer. It is important that any coating bath should be free from metals more noble than zinc. If the bath contains mercury salts they will be found in the deposit and parts so coated should not come into contact with aluminium or brass on which they cause increased corrosion or season cracking respectively.

Cadmium. The use of cadmium as a coating is relatively small since zinc is much cheaper and it is a matter of dispute which of the two gives the better protection. However cadmium has several advantages over zinc in that it is more easily soldered, has a smaller contact resistance and a higher deposition equivalent, the metal is less likely to tarnish and it is more resistant to alkalis such as washing soda. The corrosion resistances of the two metals are in general similar.

Tin. The use of tin as a protective coating is, owing to its high price, largely restricted to equipment in the food industry.

Lead. Lead coatings are largely used for protection against acid attack. Lead is more noble than iron in the galvanic series and the coating must be very dense. Pores are not dangerous since they are self blocking so that attack cannot spread out beneath the lead layer.

Chromium. Chromium coatings are mainly thin and decorative and not commonly used in immersed conditions. They are, however, valuable as hard coatings for cylinder liners and as thick deposits for the reclamation of such articles as rolls.

METAL SPRAYING

In this method the metal in more or less fluid form is sprayed from a
pistol on to the basis metal in the condition of droplets with slightly oxidised surfaces. In structure, a sprayed metal surface consists of more or
less separate particles strongly bonded one to another, the degree of frit-
ting depending on the method of application. Porous layers are suitable
for friction surfaces since they can retain lubricant but penetration by a
corrosive medium is facilitated by the oxide layer and leads to attack on
the basis metal. In some environments the pores become sealed with
corrosion product but in order to obtain maximum corrosion resistance
the layer must be compacted thermally, when diffusion occurs; mechanically, by rubbing or rolling; or chemically, by impregnation. Even after com-
pacting, the conductivity, strength and elasticity are low owing to the porosity and the presence of oxide. The hardness is generally higher than the
metal in the rolled state but this is of less importance than the resistance
to removal by abrasion.

Liquid metal, metal powder or metal wire may be used in the spraying
pistol. Liquid metal spraying is little used today because it is too cumber-
some. The powder method is used for metals difficult to melt, for hard
metals such as tungsten which are not available in the form of wire and
for spraying with non-metals such as plastics, ceramics, e.g. alumina, and
the important cermets, e.g. chromium-alumina. Among the plastics are
included polyethylene, nylon, wax, shellac and bitumen. The spraying of
easily melted metals in the form of powder has the disadvantages that
there is appreciable oxidation during spraying, loss of metal and the metal
vapour is bad for health. In the wire method the metal wire is melted in
the pistol, by gas flame or electrically, and dispersed by compressed air.
The pistol may be stationary or transportable. In general, wire-spraying
is used for metals with a melting point less than 1,600°C but even molybdenum, melting point 2,622°C, can be sprayed. By the application of a
stainless steel coating the manufacture of the whole article from this
expensive material is avoided. Coating of hard metals such as steel may
be used to reclaim worn machine parts.

The most important coatings for the protection of steel against corrosion
are zinc and aluminium. Zinc-sprayed metal may have rounded or columnar grains depending on whether powder or wire spraying is used. The
choice of method of application and the atmosphere used will affect the
amount of oxide and the porosity of the coating.

Metals sprayed with aluminium may be referred to as "metallised" and
have a dull finish but when the spraying is followed by heat treatment, e.g.
15 minutes at 900°C, they are referred to as being "aluminised" and have a
bright finish. A section of a metallised steel tube is shown in Pl. 22. The
coating has a thickness of about 6 mils and its porosity and lack of adhesion to the parent metal are self-evident. The aluminised coating shown
has, in addition, an underlying diffusion layer. There is also appreciable
decarburisation of the surface layer of the original tube.

Aluminium coatings are resistant to many atmospheric conditions and the
molten and wire-sprayed coatings give very good service in sea water.

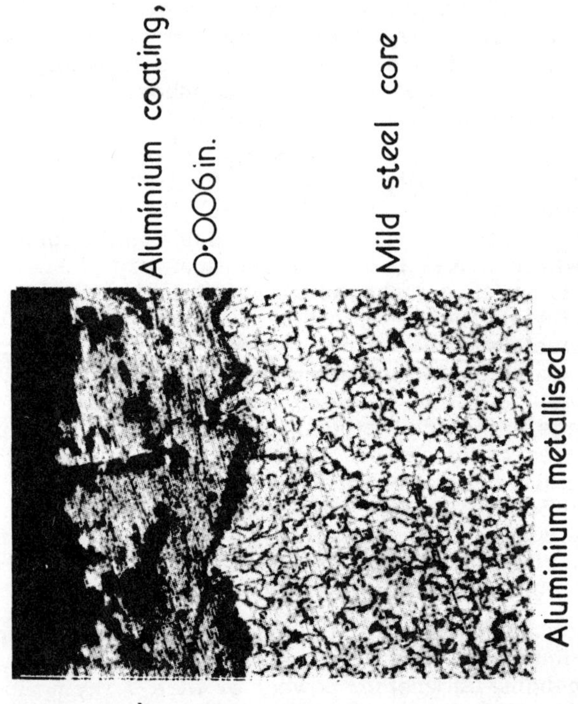

Aluminium coating, 0·006 in.
Mild steel core
Aluminium metallised

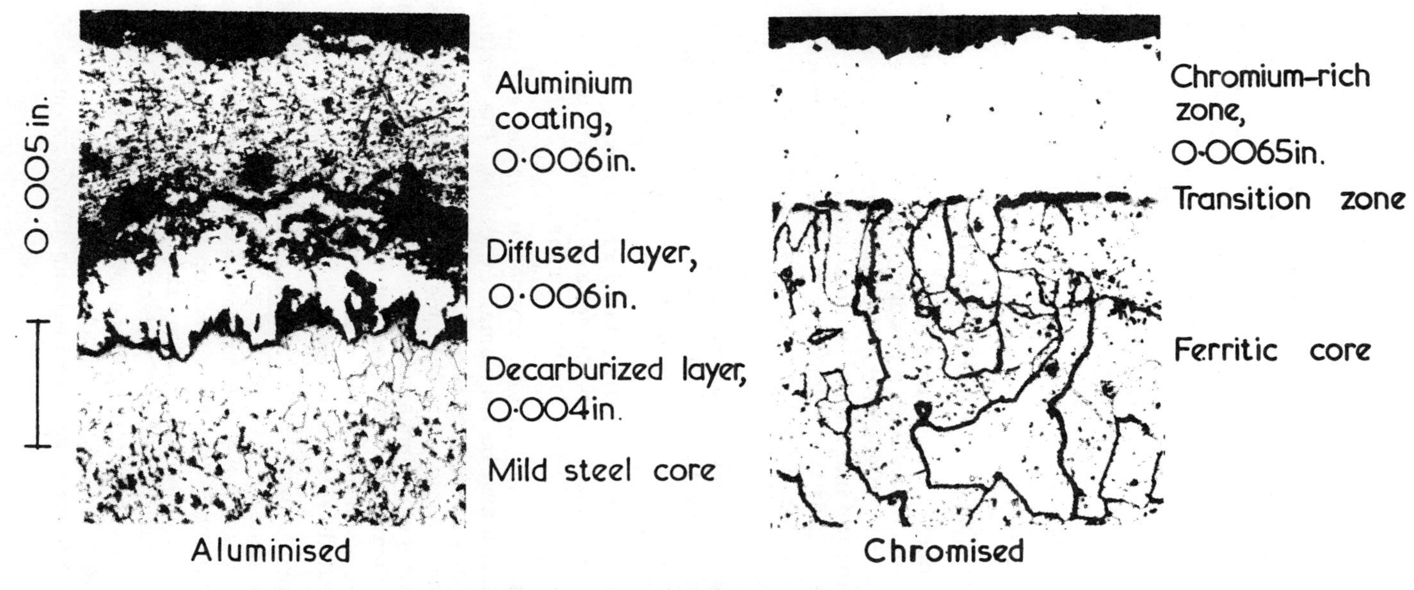

Plate 22. MICROSECTIONS OF ALUMINIUM-COATED AND CHROMISED STEEL.

They are less resistant to hard waters than to soft waters in which they are preferred to zinc. Zeolite-softened waters usually have some residual alkali and attack the coating.

Zinc is commonly used in immersed conditions and gives excellent service. The life of a coating of zinc, whether it be applied by dipping or spraying, is proportional to the thickness of the coating. To ensure starting with an adequate coating it should be at least 3 mils thick, approximately 2 oz per sq. ft. Zinc coatings are more successful in hard waters than in soft waters because the basic carbonate and hydroxide corrosion product, and the deposition of carbonate from the hard water, plug the pores and stifle further attack. There are many examples of zinc coatings lasting for more than 15 years in many natural waters but in sea water they are rarely used without further protection.

DIFFUSION AND CEMENTATION PROCESSES

In cementation processes the parent metal, usually mild steel, is heated at high temperature in contact with a powder of the coating metal. Good adhesion is obtained by virtue of the alloying of the two metals. If the coating diffuses too far into the basis metal and forms hard and brittle intermetallic compounds this rules out any subsequent treatment such as forming. This puts a limit on the permissible degree of alloying to improve adhesion.

The coating metal is either in the form of powder or one of its compounds which decomposes below the diffusion temperature. The articles are packed in the powder and heated in the absence of air, often in the presence of an inert gas, very often hydrogen. To avoid the articles baking together the metal powder is usually diluted with its oxide. It is essential that all parts of the surface of the object are in contact with the powder which must be sufficiently fine for this purpose.

The reactivity of the metal surface may be very varied and on the least active parts diffusion may be very small or absent. The surface may be evenly activated by grit or sand blasting or by the addition of an activating agent to the powder. Ammonium chloride is often used up to a concentration of 10 to 15 per cent. Undesirable decomposition of the ammonia may lead to nitride formation and interfere with the diffusion processes. This gives less trouble when the carbon content of the iron is low or when the surface to be treated is decarburised.

The best known processes are <u>sherardising</u>, diffusion coating with zinc, and <u>calorising</u>, diffusion coating with aluminium. Sherardising is used for small articles only since the coating tends to be fissured. Its main advantage over dipping is that the operation can be carried out at a temperature about 100°C lower than the melting point of zinc. On the other hand, coating with aluminium has a wider application since the coating performs best against high temperature oxidation. A mixture of aluminium, aluminium oxide, carbon and ammonium hydroxide is often used. If pure aluminium is replaced with an alloy containing 50 per cent iron a higher temperature and shorter time of diffusion can be used. Material is saved if the surface is coated with aluminium by dipping or spraying and then heated. The sprayed layer is very porous and must be protected from oxidation during heat-treatment by a coating of bitumen, water glass or borax solution.

Diffusion methods are used to a much smaller extent for coating with other metals including chromium, copper, tungsten, beryllium, tantalum, silicon, manganese, vanadium and titanium. These metals are more usually applied by the vaporisation method (see later).

Sometimes the surface of a metal may be hardened by heating a deposited coating, e.g. by diffusing copper into aluminium. This method may also be used to prepare alloys which cannot be deposited electrolytically because of the different deposition potential of the constituents. Coating with a nickel-chromium alloy is an example of this. The separate metals are deposited in alternate layers on the surface and then by thermal treatment an alloy is formed which is corrosion resistant.

VAPORISATION METHODS

In these methods, metal compounds, usually halogenides, are thermally decomposed on the metal surface. One of the most important coatings prepared in this way is that by chromium diffusion, <u>chromising</u>. Dry chromous chloride vapour mixed with an inert gas, usually hydrogen, is passed over the metal which in the case of iron is at a temperature in the range 900 to 1,000°C. To avoid corrosion, the hydrogen chloride formed must be removed rapidly from the reaction space. The deposited chromium diffuses into the iron to give an alloy with a high corrosion resistance. The mechanical properties of chromised parts may be impaired by excessive grain growth and by carbide formation. Since a high carbon content hinders diffusion it is desirable to use special steels containing titanium which fixes the carbon as titanium carbide. The final coating consists of a layer of columnar ferrite rich in chromium (21 to 25 per cent) on the outside, a narrow transition zone changing to austenite at about 12 per cent chromium and, underneath, the original steel (Pl. 22).

An alternative method is to pack the objects to be coated in chromium metal powder and to pass hydrofluoric gas through the mixture.

The gaseous methods are particularly desirable for depositing metals or compounds such as carbide, boride, nitride and silicide which cannot be deposited by other means. The temperatures involved are in the range 1,000 to 1,400°C.

MECHANICAL CLADDING

The advantage of this method over the others is that the coating is completely free from porosity and may be of any desired thickness. In one method the coating metal is cast around a steel billet in a mould and the composite billet is then heated and rolled in a non-oxidising atmosphere.

The method most used is roll-bonding in which the coating metal at the desired thickness is rolled onto the basis metal at the welding temperature. Both metals must be clean and free from oxide. In another method, cladding is made under heat and pressure by the use of an intermediate layer of bonding material.

The basis material must have a good welding ability and a low carbon content. Owing to its mechanical properties mild steel is generally used but in particular cases may be replaced by special steels such as stainless and acid-resistant steels. The coating metals are mainly copper, nickel and aluminium, but more recently alloys with high corrosion resistance such as stainless steel have been used. A well known example of cladding is that in which an aluminium alloy with good mechanical properties is clad with a thin sheet of pure aluminium which has superior corrosion resistance.

This principle has been extended and it is now possible to obtain many duplex sheets or tubes which can solve difficulties in which mechanical properties and corrosion resistance are not compatible or in cases where the environments on either side of a tube wall are different or of differing corrosion characteristics.

INORGANIC COATINGS

A number of metals such as chromium and aluminium rely for their corrosion resistance on a surface layer of oxide or hydroxide. The films formed naturally are very thin, 5 to 150 Å thick, and a number of methods have been developed for improving the protective value of such films, particularly on aluminium, so that the corrosion resistance is higher. One obvious way is to increase the thickness of the oxide layer by raising the temperature and increasing the amount of oxidation.

Methods evolved for improving the corrosion resistance of the surface layer include treatment in solutions of oxidants such as chromate and permanganate or in solutions to produce fluoride or phosphate layers which are used as a pretreatment for painting.

OXIDES

Much more resistant and thicker oxide layers may be formed by electrolytic anodic oxidation by alternating or direct current in acid solution. To produce thick oxide layers three conditions are necessary; the solution must be able to attack the metal anode to some extent, formation of capillaries in the oxide must take place and the solubility of the layer formed in the electrolyte must be low. The layer forms from the metal outwards and pores are necessary for the electrolyte to reach the metal. However, since the electrolyte is able to dissolve the oxide layer, the temperature must be adjusted to make the rate of dissolution slower than the rate of formation. Since in the capillaries there is probably a fairly high current density, heating will take place leading to the decomposition of aluminium hydroxide to the very hard modification of γ-alumina. Consequently the layer is not homogeneous but is hard and dense next to the metal and soft and less dense on the outside. The importance of this type of surface treatment is that it can be applied to an article of any size.

Chromic, sulphuric or oxalic acids are generally used and various film thicknesses may be obtained, a typical one being 0.1 mil so the films are at least 200 times thicker than the air-formed film.

The porosity of the layers may be reduced and their corrosion resistance improved by sealing the coating. This may be carried out by immersion in boiling water when the anhydrous γ-alumina is converted into the crystalline hydrated oxide Boehmite, $Al_2O_3.H_2O$, which has greater volume and blocks the pores. Solutions of chromates, silicates and borates may also be used for sealing as well as impregnation with waxes, oils, greases and silicones. The sealing method can also incorporate organic or inorganic colouring matter which is very useful for decorative purposes.

Oxide films may also be formed on magnesium by similar anodic treatment in strongly alkaline solutions containing fluoride. The layers on magnesium are, however, soft and can only be used as a basis for paints.

Oxidation of titanium can be carried out by anodic treatment in solutions of potassium hydroxide, borax or sulphuric acid. The corrosion resistance of the layer formed depends on whether the water contains reducing or oxidising agents. In reducing conditions both metal and oxide will be attacked but in oxidising or neutral media the oxide-coated metal is resistant.

Oxide coatings on steel can be produced by oxidation at high temperatures in air or by immersion in hot concentrated alkali solutions containing oxidising agents such as persulphates, nitrates or chlorates. Such coatings consist mainly of magnetite. One commercial method of treatment of steel for use under immersed conditions, specifically in domestic boilers, is Bower-Barffing. The steel is heated first in air and then in super-heated steam, when a layer of mixed iron oxides, Fe_3O_4 and Fe_2O_3, is produced. The heating is then continued in producer gas which converts the whole coating to magnetite, Fe_3O_4. This type of coating gives good protection if it is continuous.

PHOSPHATE

The corrosion protection given by both phosphate and chromate coatings is very small and their main use is to give a good key for paints. Phosphate coatings are very porous and chromate coatings, although dense, are very thin.

In phosphate coatings we are concerned with salt deposition. As phosphoric acid is neutralised by metal dissolution, the successive replacement of the three hydrogen atoms in the acid, H_3PO_4, by metal produces primary, secondary and tertiary phosphates which have solubilities decreasing in the same order. If iron is dipped in phosphoric acid some of the metal first dissolves to form the primary phosphate which subsequently hydrolyses to the less soluble secondary and tertiary salts. If the temperature and conditions are right these products are deposited on the iron.

The use of a simple phosphoric acid bath has several disadvantages; long immersion times are required, the composition of the bath changes rapidly and large amounts of sludge are thrown down, and the protective value of the phosphate coating is variable. To overcome these difficulties many addition agents have been employed including primary phosphates of metals such as manganese and zinc, and oxidising agents, such as nitrates, nitrites or chlorates.

A phosphating bath usually contains phosphoric acid and the primary salts of manganese, zinc or iron together with an oxidant.

Phosphating is usually carried out at elevated temperature but can be carried out at ambient temperature if an acid sodium fluoride buffer is added to the phosphate bath to keep the pH value at about 2.7. Phosphate layers may also be produced electrolytically.

A large number of phosphating treatments exist under a variety of trade names,e.g. "Bonderising", "Coslettising", "Parkerising" and "Walterising" etc.

As previously remarked, the layers are very porous and cannot be considered adequate corrosion protection, being usually employed as a basis for paints. The metal to be coated and the surface treatment can both influence the nucleation and crystal growth but this influence is not very marked in modern baths where the dip time is short.

Deposits may be made on iron, zinc, cadmium and aluminium. The layer thickness and treatment in the bath may be adjusted according to the objective, and representative thicknesses of commercial coatings are in the range 0.04 to 0.3 mil. Resistance to rusting can be improved by a variety of secondary treatments including sealing or impregnating of the porous layer with oils or waxes especially if they contain corrosion inhibitors.

CHROMATE

On dipping a metal in a bath containing chromic acid or an acid solution of dichromate a greenish-yellow layer of basic chromate with a thickness of 0.02 to 0.04 mil is formed on the metal. Such coatings can be formed on cadmium, copper and bronze but the most important use is on zinc, aluminium and magnesium. The chromate layer formed protects the metal against spotting and staining, e.g. by perspiration, and also to some extent against atmospheric corrosion. Chromate films are generally used as a basis or key for paint, particularly where great reliance is placed on the protection of the metal.

The protective value of chromate coatings is quite small but in many cases it can increase the corrosion resistance of a zinc surface 15 times. For good protection it should also be given a coat of varnish.

TEMPORARY PROTECTIVES

As the name implies these coatings are primarily of use only in storage
or transit, and have little or no value under immersed conditions. It is
usually desirable that there shall be no deterioration of a metal surface
during the very long time that may elapse between manufacture and final
coating or painting of the finished structure and in such cases a temporary
protective must be used. This may consist of a straight solution of lanolin
or petroleum jelly in a solvent, a hard pigmented coating, a transparent
lacquer or synthetic resin, silicone grease or a more simple grease or oil.
For long periods, a zinc dust-epoxy resin primer has shown promise. The
coatings can be applied by dipping, brushing or spraying. For a limited
period of storage steel can be protected with a film of boiled oil. Fluids
which have the property of displacing water from a wet surface and then
evaporating to leave a protective film are extremely useful. Vapour
phase inhibitors (see Idle Boilers, Chap. IX) can be sealed inside hollow
components or enclosed with articles in an air-tight container, e.g. a
plastic bag, for the protection of surfaces.

PAINTS

A paint is basically an insoluble pigment dispersed in a fluid medium,
which is usually organic and decides the physical and chemical properties
of the paint, and is modified by driers and thinners, which are volatile
compounds and are added to control the viscosity both for ease of manufacture and subsequent application. There is an infinite number of combinations of these constituents and it would be impossible to deal adequately with all of them in a few pages. Here only a broad outline of the formula-
lation and application of paints will be discussed.

There are three main factors to be considered in the use of paints: adequate preparation of a suitable surface to receive the paint, the correct
choice of paint system for a given application, and correct application of
the paint. We have already considered in some detail methods of cleaning
the surface from grease, dirt, rust and scale and if such a surface is to
be finally painted we must ensure that an adequate 'key' for the paint is
also obtained.

Before a paint can be specified it is essential to consider carefully all the
conditions of service; the best paint system in one case will not necessarily be applicable in another. Thus a paint that behaves satisfactorily in
sea water may blister badly in fresh water owing to the difference in os-
motic pressure between the two media. Again a paint which is satisfactory
in fresh water may be unsuitable in sea water where fouling has to be
considered. The pigment used is very important, particularly in the
primer, and an unsuitable choice of primer can lead to oxidation of the
painted metal leading to severe localised attack when the paint breaks
down.

A paint system usually consists of a primer, selected for adhesion and
protection of the substrate, undercoats which have a high pigment content

and a low gloss, and a final coat chosen for protection and finish. In marine applications an additional coat of an antifouling paint is necessary to prevent the growth on the surface of marine organisms. In addition a number of fillers and putties are used to fill any crevices or holes and give the final paint system a good smooth finish.

PAINT COMPOSITION

Vehicle. The vehicle is the most important component of a paint and most commonly is a natural oil such as linseed or tung oil. On exposure to the air these 'drying oils' oxidise and polymerise to solids, a process which can be accelerated by pretreatment of the oil or by adding small amounts of driers. Raw linseed oil is very slow in drying, is lacking in gloss and does not flow sufficiently. Hence linseed and tung oils are treated in various ways to improve their properties. Boiled oils are those heated with a drier, stand oils are oils heated with the exclusion of air which brings about a degree of polymerisation, and blown oils are oils which are heated without a drier but with air blown through them producing a material rich in polar groups such as hydroxyl.

Synthetic resins are often used nowadays as vehicles or components of vehicles, particularly for continuous contact with water and use at higher temperatures. The resins may be dissolved in a volatile solvent, which evaporates to leave the resin, or they may be liquid and polymerise through the action of heat or by action of a suitable catalyst. The synthetic resins include phenol-formaldehyde formulations, which can withstand water at the boiling point or slightly higher, and are used as multiple coats baked on for resisting a variety of corrosive media. Silicone resins may also be used at elevated temperatures.

Vinyl resins are resistant to penetration by water and their resistance to alkalis makes them suitable for painting structures which are protected cathodically. Linseed and tung oil paints on the other hand are quickly saponified and disintegrated by the alkali formed at the cathode. Epoxide resins are also resistant to alkali and adhere well to metal surfaces. They are the basis of plastic mixtures which, with an addition of a suitable catalyst, solidify very quickly making them useful for sealing leaks in ferrous or non-ferrous piping. Other vehicles include alkyd or amino resins, polyurethanes, chlorinated rubber and nitrocellulose.

Ordinary paints with a simple linseed-tung oil vehicle are not suitable for structures immersed in water and have a lifetime of only one year or less at ambient temperature and a still shorter period in hot water. Better protection of up to several years at ordinary temperatures is obtained from a multi-coat system of synthetic vehicle paint but since this is expensive many applications for fresh or sea water make use of thick coal tar coatings.

Driers. These are soaps, usually naphthenates or linoleates of manganese, lead or cobalt used singly or together. The amount required is very small, about 0.05 per cent of the lead salt and 0.003 per cent of the manganese.

Thinners. For oil-bound paints white spirit is now more commonly used but turpentine was preferred at one time. The latter is more expensive than white spirit but is a better solvent for the lead soaps formed from white lead. The synthetic resin paints each need a particular thinner which must be compatible with the type of resin medium.

Pigments. Probably the best known and most used pigment is red lead, Pb_3O_4, which is used in primers for steel work but is likely to accelerate the corrosion of non-ferrous metals. Calcium plumbate, Ca_2PbO_4, is unique in its adhesion to newly galvanised surfaces but it is less effective than red lead on steel. Zinc chromate may be used for both ferrous and non-ferrous metals and is the inhibitive constituent of etch primers. Zinc powder find an important use as a pigment in the protection of steel and, provided about 95 per cent by weight of zinc is present, it is able to protect steel cathodically. Iron oxide, Fe_2O_3, and titania, TiO_2, do not have any inhibiting properties but are able to decrease the permeability of the paint.

MECHANISM OF PROTECTIVE ACTION

A paint film can affect the electrochemical corrosion reaction by inhibiting the cathodic or anodic reaction or by providing a high resistive path between the anode and cathode. In order to influence the cathodic reduction of oxygen the paint film must either hinder electron transfer or the transport of oxygen and water. Although there will be no electron transfer, since the conductivity of paint films is ionic, more than enough water and oxygen can usually pass through to maintain the corrosion rate of steel at the unpainted value.

The anodic reaction, i.e. the dissolution of metal ions, will be prevented if the metal is cathodically protected or inhibited by the paint film. The metal will be protected cathodically if the primer contains a less noble metallic pigment, the particles of which are in contact one with another and with the metal. The only commercially available pigment of this type is zinc dust which will protect iron if the zinc content of the dried film is about 95 per cent by weight. The paints are porous but once the pores become blocked with hydroxides and carbonates of zinc, calcium and magnesium it is impervious and metallic contact is no longer necessary.

Anodic solution may be prevented if the pigment is itself a soluble inhibitor or if it reacts with other constituents of the paint to form an inhibitor. Basic lead carbonate or sulphate, red lead and zinc oxide form lead or zinc soaps which degrade to form inhibitive substances of unknown composition.

Of the soluble inhibitive pigments the most important are the metallic chromates. Zinc chromate and basic zinc chromate or zinc tetrahydroxychromate are soluble enough to render water non-corrosive and thus protect the metal under the paint. Zinc chromate is usually used in a synthetic resin vehicle such as an alkyd resin, in which it performs better than in linseed-tung oil.

The diffusion path through the paint may be increased by using pigments which are in the shape of flakes, e.g. aluminium powder or flaky haematite, and are orientated parallel to the metal surface, e.g. by brushing.

Many protective paints do not contain inhibitive pigments and in this case the corrosion is prevented by the high electrical resistance which limits the corrosion current. This can operate by opposing the passage of anions, such as chloride and hydroxide, making it difficult for electrolyte to reach the metal. In general, paint films which have a high electrical resistance give better protection against corrosion than those with a low resistance.

MECHANISM OF PAINT BREAKDOWN

The incorporation in the paint film of ion-producing materials from pigment or vehicle reduces the resistance and leads to breakdown. This can be eliminated by the selection of suitable raw materials by the paint manufacturer.

In a correctly applied film of paint the breakdown is usually associated with the nature of the paint film and the water or electrolyte outside the film. Owing to their constitution, paint films of linseed and tung oils, as well as polystyrene and other polymers, have a selective permeability. The rate of diffusion of ions is much smaller than the rate of diffusion of water and oxygen and is proportional to the conductivity, i.e. inversely proportional to the resistance. The resistance is controlled by the uptake of water or by the exchange of ions in the solution with hydrogen ions in the paint. Thus there is a lower resistance in distilled water than in sea water.

Blistering and swelling can be caused by steam, water vapour or condensate which leads to corrosion under the paint film. Under tropical or very humid conditions mould growth, mildew or bacterial attack can occur. This can be remedied, when it occurs, by washing the infected surfaces with an antiseptic but it is better to incorporate some toxic material in the finishing coat.

APPLICATION OF PAINTS

The desirable conditions for application of paints include adequate surface preparation, an appropriate method of application under suitable climatic conditions, the use of suitable paints, and suitable environment and design.

Surface Preparation. A number of methods of removal of rust and mill-scale, grease and dirt have already been considered. After preparation of the surface it is desirable to apply the paint coating immediately since, particularly after mechanical cleaning, the surface is left in a condition highly susceptible to corrosion. On the other hand some people prefer to let rust form on a steel surface which, in the absence of mill-scale, is fairly uniform and is largely removed by wire-brushing. If the metal has been phosphated, painting may be delayed for a short time. The condition of shop-applied paints should be examined on site and at any areas where the coating has been damaged, or where fitting operations have been carried out, e.g. cut edges and sites of welds and rivets, the paint must be made good.

When repainting it is essential that all contaminants including loose paint should be removed from the surface.

Climatic Conditions. It is bad practice to apply paints to surfaces carrying electrolytes such as films of moisture caused by a drop in temperature or deposited from rain, fog or dew. These not only affect adhesion and reduce the paint life but, due to the prolonged drying time, dirt and dust can become attached to the painted surface. In ports, docks and coastal areas up to several miles inland, the difficulties are increased by the saline atmosphere and erosion from blown sand. Again, in industrial atmospheres where the sulphur dioxide content is high the rain water is acid and may have a pH value as low as 3.0.

Wherever possible, paint application should be carried out under dry conditions. Paints applied in the winter often give an inferior performance to those applied in the summer. This is attributed to the ferrous sulphate and other salts trapped beneath the paint. These lead to the rapid formation of voluminous rust which forces the paint outwards and eventually disrupts it.

In the spring and early summer, rain is relatively pure and any salts are washed away. The hydrated ferric oxide remaining has little or no effect on paint behaviour compared to rust formed in the presence of salt solution which leads to rapid breakdown. If steel has to be painted without complete removal of rust, red lead paint gives a reasonable performance but calcium plumbate an even better one.

Choice of Paint. The paint must have certain basic qualities such as good spreading power, elasticity, resistance to the environment and compatibility with the metal. Adhesion is a primary requirement and, while it may not be a problem with a well-prepared steel it can be troublesome with some metals particularly when they have smooth surfaces. Etch primers have been developed especially for this type of application. It should be remembered that any given paint, properly applied, has a life dependent on its thickness.

Alkali Resistance. All paints for marine use must be resistant, in some measure, to alkali. Even in the normal corrosion process alkali is formed at the cathodic points and this is increased under paint films which are pigmented with zinc or aluminium and may be particularly high when cathodic protection is applied externally. In the latter case it is absolutely essential that an alkali-resistant paint is used around the anodes. The standard red lead/linseed oil type of paint for structural steel is not of great use when immersed in sea water. For the painting of ship' bottoms the pigment is often complex and may include aluminium powder while the vehicle may be linseed oil or linseed/tung oil modified by the addition of phenolic or other resins.

Corrosion-Promoting Pigments. There is a real danger that the pigment
in the paint may actually promote corrosion particularly if incorporated in
the first coat or primer. Common ones which can give trouble if applied to
certain metals are the red oxides or iron, gypsum, ochre, graphite and
lamp black. Anti-fouling paints containing copper and mercury should
never be used in direct contact with steel. The mercury compounds used
as fungicides in sea water also have a tendency to corrode aluminium and
its alloys.

Repainting. Economically a choice has to be made regarding the correct
time for repainting because if it is delayed too long it may be necessary
to clean all the structure down to the bare metal. It has been recommended
that the best time is when only 0.2 to 0.5 per cent of the surface shows
evidence of rust.

SPECIAL PAINTS

Primers. A zinc-rich primer may be applied directly to a lightly corroded
steel surface when any rust is reduced to magnetite or metallic iron. The
mechanism of protection by such paint has already been discussed. These
paints are extremely useful and, although in no way considered to be as
good as galvanised or sprayed coatings, they provide a reasonable alternative when these processes are impracticable and are valuable for touching
up and repainting.
Zinc chromate is widely used as an inhibitive primer but is not so effective on zinc and gives poor adhesion when zinc phosphate is present.
Calcium plumbate may be applied directly to zinc or galvanised steel
without prior treatment. All inhibitive primers should be applied directly
to the metal surface.

The primers which have found wide application are those variously known
as etch primers, wash primers or wash coats. They combine the advantage
of a phosphoric acid wash and a zinc chromate pigment. They are two-
solution primers which are mixed immediately before use, one contains
zinc tetroxychromate and polyvinylbutyral resin in alcohol and the other
phosphoric acid in alcohol with a small amount of water. The resulting
resinous film is very thin and, although it does not replace any of the paint
coats, it adheres well to the metal surface and forms a particularly good
inhibitive bond for the paint. Wash primers are effective on steel and
most non-ferrous metals and are widely used on aluminium and its alloys.
They are also effective on steel-aluminium couples.

Epoxy-Resin Paints. Epoxy-resins, which have found great favour in recent
years as paints and coatings, are a large class of organic polymers ranging from viscous liquids to solids with melting points of about 140°C.
There are basically four types of epoxy-resin paints:-

1. Those made from epoxide esters which dry at ordinary temperatures
    and are similar to ordinary paints.

2.   The cold-setting coatings which harden at ordinary temperatures by chemical reaction. These are two-pack systems the resin and the catalyst, amine or amine-adduct, being mixed in simple proportion immediately before use.

3.   Low temperature coatings which are stoved at 120 to 150°C.

4.   High temperature coatings which are stoved at 175 to 205°C.

The first type of epoxy-resin paint has no particular advantages over the more conventional underwater paints. The second type is more protective but the pre-mixing may be considered an inconvenience and furthermore, to prevent excessive waste it is necessary to estimate accurately the amount required.

The coatings are resistant to alkali and are finding increasing use for patch painting around anodes used for cathodic protection. Epoxy resins (and vinyl and isocyanate resins) are also used for the protection of cargo and ballast tanks. A total thickness of 4 to 6 mils is required and epoxy coal-tar pitch and solventless epoxies applied thickly in one coat appear to be promising.

Cold-setting epoxies are not to be recommended for the interior of tanks used to contain water for drinking or washing since there is a danger of a continual leaching of small amounts of excess catalyst from the coating. In these circumstances an epoxy resin coating of the stoving type must be used.

Cementiferous Paints. One of the difficulties of paint application, particularly in the case of ships, is that it often has to be carried out under humid or even wet conditions. This difficulty may be overcome by the use of paint formulations based on inorganic media such as sodium, lead and ethyl silicates, oxychloride cements, phosphates and butyl titanate. The pigment used is usually zinc dust and in a silicate vehicle both zincates and silicic acid occur in the film. These paints are resistant to water and all the usual petroleum cargoes but are not resistant to strong acid or alkali because of the high zinc content. While in organic media it is necessary to have sufficient zinc to give metallic contact between particles and basis metal, this does not apply in inorganic media, such as silicate, which react with the zinc, making it chemically part of the coating.

Bituminous coatings. There is sometimes some confusion in the use of the term 'bituminous' which merely means 'resembling bitumen'. Bitumen and asphalt are non-crystalline solid or semi-solid substances which are essentially hydrocarbons. They are found naturally or are produced as residues in the distillation of crude petroleum. Bituminous coatings will also include coal tar and pitch which are the residues from the carbonisation of coal.

The coatings may be applied cold or hot depending on formulation and they have been used extensively for lining water pipes and tanks, the pipes being coated usually by a dipping or by a centrifugal method. 'Angus Smith' compound is synonomous with a thick hot coal-tar coating applied to pipes. The protection obtained depends on the thickness of the coating and it is difficult to obtain one free from pinholes unless it is approaching 0. 25 in. in thickness. The bituminous paints have been modified by the addition of pigments and drying oils and the coal tar enamels have been used extensively in marine conditions.

Graphite Paints. There appears to be only one type of paint that is regularly used for higher temperatures in immersed conditions such as those in low-pressure boilers. This paint is based on graphite and is reputed not only to prevent corrosion of boiler shells but also to prevent the deposition of a hard adherent scale which is difficult to remove. The paint is of interest because a graphite-steel combination is dangerous at ordinary temperatures since graphite is very cathodic to steel. It therefore seems likely that under boiler conditions their relative positions in the galvanic series are altered.

Anti-Fouling Paints. Fouling can cause a 50 per cent increase in fuel consumption on ships because of the drag caused by the increase in surface roughness. It can be prevented by applying anti-fouling paints which operate by slowly releasing into the water chemicals, usually copper or mercury salts, which are poisonous to living organisms. The life of many such paints is limited to 9 to 12 months but special types are available which last for about 3 years. An anti-fouling paint based on cuprous oxide/ sodium silicate has given a life of 2. 5 years in sub-tropical waters.

OTHER COATINGS

Vitreous Enamels and Glass. These are some of the oldest established coatings and are used extensively on tanks and domestic utensils but need a high temperature (650-850°C) to fuse the silicate or borosilicate glass coatings. These enamels are very resistant but accidental damage often causes chipping which renders the coating useless.

Rubber. This is widely used for pipes and tanks and its resistance to waters is excellent. Some success has been reported in the coating of ships' propellers to combat cavitation.

Plastics. All the plastics which can be coated by brushing, dipping or spraying are used for protecting metals. Provided the coatings are adherent and non-porous the resistance is that of the plastic itself. Most of them are reasonably resistant to waters but only withstand moderate temperatures and are often soft enough to be easily damaged.

Polyvinylchloride (PVC) and polyethylene can be used up to about 80°C and, as they are easily welded, can be used for lining tanks. The softening temperatures can sometimes be exceeded without impairing the efficacy of the coating provided it is under no external stress.

Polytetrafluorethylene (PTFE) and polytrifluorchlorethylene (PTFCE) withstand higher temperatures but are more difficult to apply. The thermoplastic PTFCE can be dispersed in an organic medium and applied by brush but must then be sintered at about 290°C and even then several coats must be applied to ensure that there is no porosity. PTFE is a much more difficult substance to handle and, while its properties of chemical resistance are outstanding, by its very nature it creates problems of adhesion. Difficulties of extrusion, moulding and application make it expensive as a lining.

Concrete. Concrete is still successfully used for lining large diameter mains for transporting corrosive waters. It may be used for hot water providing there is no sudden change of temperature which can fracture the coating. It appears to become coated with a film of ferrous hydroxide which seals the pores and renders it more resistant.

From the foregoing it can be appreciated that the choice of a protective coating can be bewildering. Many are specialist coatings and may be used for small particular jobs where the cost of the coating is of minor importance even though it may be high enough to preclude its use for larger plant. Whatever the coating used, it can be assumed that while intrinsic properties such as inhibition are of value, the most important properties are those of adhesion, lack of porosity and resistance to the environment. These properties will materially depend on the preparation and pretreatment of the metal surface and the coating thickness. The choice is difficult and it must be reiterated that the best results will be obtained only if expert advice is sought.

Chapter XI

CATHODIC PROTECTION

Cathodic protection is the application of a counter-current sufficiently
large to neutralise the currents responsible for corrosion. Under these
conditions the whole of the surface which was previously corroding is now
a cathode and the anodic process occurs on an auxiliary anode. The counter current may be supplied by more electro-negative metals, usually
magnesium, aluminium or zinc which undergo "sacrificial attack". Alternatively one may use an impressed current from a generator using auxiliary anodes of iron, steel, graphite, lead or platinised titanium. This is
illustrated schematically in Fig. 40. A and C represent the anodes and
cathodes of a corroding structure. If we now use an auxiliary anode and
impress sufficient current, the anodes and cathodes of the structure will
be polarised to the same electronegative potential and the whole structure
will become the cathode, C′ C′, to the auxiliary anode and will therefore,
be protected.

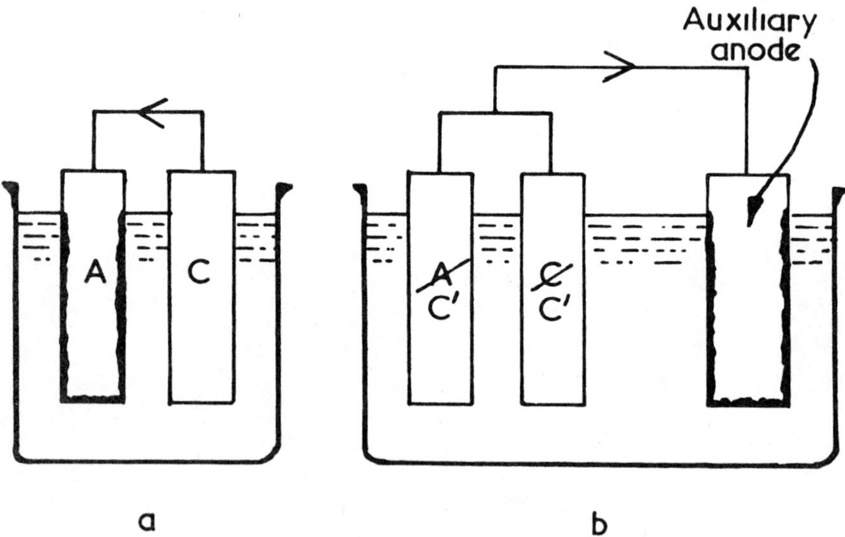

Fig. 40. DIAGRAM ILLUSTRATING THE PRINCIPLE OF
CATHODIC PROTECTION.

    a. structure without cathodic protection
    b. structure with cathodic protection

This method of protection was applied as long ago as 1824 by Sir Humphrey Davy who used zinc sacrificial anodes to combat the corrosion of the copper sheathing of warships. Although use has been made of zinc protector blocks since that date it is only over the last twenty years or so that cathodic protection has had a much wider application. This is the outcome of a large volume of research and a better understanding of the underlying principles. Although normally applied to steel structures it has also been used for the protection of lead, aluminium and copper alloys.

The increase in alkalinity and pH on the cathodic surface of the structure has very important secondary effects. In sea water and many other natural waters it causes the deposition of insoluble calcium carbonate and magnesium hydroxide which form a protective scale. On the other hand, the alkali causes saponification of oil-bound paints leading to softening and blistering. Furthermore it will attack metals such as aluminium and lead and when protecting these metals cathodically the current must be kept under close control.

When applying cathodic protection it is essential to avoid certain undesirable side effects. Other metal structures in the vicinity may undergo increased attack owing to stray currents. When applied to copper or its alloys fouling by marine organisms may result since the small corrosion rate necessary to prevent fouling is eliminated. If, when protecting systems internally, the current is excessive, oxygen will be liberated at the anode and increase the corrosion of parts remote from the protected area. In sea water excessive current may cause the evolution of chlorine gas. Lastly, high current densities give rise to the possibility of sparks which are dangerous in the presence of inflammable liquids and vapours.

METHODS OF PROTECTION

SACRIFICIAL METALS

As will be obvious from their position in the galvanic series, magnesium, zinc and aluminium can all be used to varying extents for cathodic protection. These metals have a tendency either to self corrosion or to the formation of protective films which reduces their efficiency as anodes. This may be avoided by using high purity materials or by making suitable alloying additions.

The self-corrosion of magnesium is very dependent on traces of copper, lead and tin, and particularly iron and nickel which should not be present in amounts greater than 30 ppm. The anode must either be made from electrolytic magnesium of high purity or, more commonly, commercial magnesium to which alloying elements are added to counteract the effect of impurities. A suitable alloy of magnesium contains 6 per cent aluminium, 3 per cent zinc and 0.2 per cent manganese.

With zinc the impurities, particularly iron, must be restricted to very low limits or the metal modified by alloy additions. Zinc of 99.99 per cent purity is recommended, with an iron content of not greater than 15

ppm. The effect of iron may be neutralised by alloying with aluminium and
zinc anodes containing 1.0 per cent aluminium remain active even when
the iron content is as high as 200 ppm. The addition of silicon reinforces
the action of the aluminium in reducing grain size which gives a more
even dissolution in contrast to the large pits which develop on pure zinc.

When the conductivity of the water is high, e.g. sea water, an anode such as
zinc giving a smaller potential difference is more economic and efficient
than magnesium and has the advantage of reducing damage to paint work.
Zinc anodes are largely self-controlling since with a low driving potential
the resistance is controlled by the resistance at the surface of the protected metal, i.e. the cathode, and the current output varies with the requirements of the cathode. Apart from these advantages, zinc gives a more
even current distribution and a long life with the minimum of maintenance.
If the water is of high resistivity however, the voltage obtained is inadequate and magnesium must be used.

Unlike zinc and magnesium, aluminium in the pure state is self-passiva-
ting and develops an oxide film which restricts its activity as an anode.
An alloy containing 5 per cent zinc is preferable but even this has a tendency to gradual polarisation and a more recent alloy containing 0.2 per
cent tin is better. Another useful alloy contains both zinc and calcium,
present as Al_3Ca. Passivation of such an anode is prevented by the formation of calcium hydroxide as alkaline corrosion product. Aluminium is
not so widely used as magnesium and zinc as an anode material and is not
used where there is a fire risk since, like magnesium, aluminium is in-
cendive.

IMPRESSED CURRENT

With the big advances in the development of transformer-rectifier sets
and automatic control widespread use may now be made of impressed
current in cathodic protection. It is now usual practice to use such
methods if possible although the cost is two to three times that of the
conventional systems. Economic considerations limit their application to
the larger schemes. The use of anodes of scrap steel and cast iron, e.g.
old railway lines, cylinder blocks and marine buoys, has the advantage of
cheapness, simplicity and availability, but these need regular replacement
and more permanent anodes have consequently been developed.

One of the first permanent or non-sacrificial anodes to be used was graphite but the application of this material is limited by its brittleness and
spalling caused by surface gassing. The behaviour may be improved by
impregnation with wax or linseed oil and resin-impregnated graphite
anodes have been used at temperatures up to 170°C. In open water their
use is prevented by a number of problems and in heat exchangers trouble
can arise from detached carbon settling on the tubes and setting up galvanic cells.

As an anode material graphite has been largely superseded by a silicon-
iron alloy containing 14.5 per cent silicon which may be used in both
fresh and salt water and may replace aluminium and graphite in hot water.

The alloy is brittle and should be adequately supported but it has the advantage that it may be used at high current densities. The pitting of silicon-iron alloys in sea water can be reduced by the addition of 3 per cent molybdenum and sometimes chromium.

In its turn silicon iron is being superseded by platinised titanium or tantalum. Titanium, although having many desirable mechanical properties, is not suitable for an anode since the extremely adherent layer of oxide on its surface has a high resistance and will not pass sufficient current. However by depositing on it a layer of platinum only 0.1 mil thick current flow may take place, the combination acting as a very efficient permanent anode which is robust and light. Titanium is a relatively poor conductor compared with some other metals but this may be overcome by adequate cross-sectional area or the use of copper coring, e.g. copper-cored platinised titanium rod. All connections and cables are best encased in unplasticised PVC to render them water-tight. The workability of titanium enables anodes of any desired shape to be designed and manufactured. Platinised titanium will operate at current densities as high as 100 A per sq.ft. which allows the anode to be made correspondingly smaller. The titanium basis metal is resistant to attack under anodic conditions provided that its potential does not exceed 12 to 14 volts and the high electrical resistant film of oxide formed ensures that the current flows only from the platinised areas. Such an electrode will operate in fresh waters of low conductivity, e.g. 25 to 200 ohm cms. There is some reason to suspect that a certain dissolution of titanium may occur if the d.c. contains a.c. components but this effect decreases with increasing frequency and trouble should only arise with poor rectification.

Tantalum is more expensive than titanium but can be used up to about 60 volts on the unplatinised areas instead of the figure of about 12 volts for titanium.

Platinum and platinum alloys are more expensive than platinised titanium but offer advantages if the anodes have to be extremely small and the current density has to be high.

Lead and lead alloys, e.g. one containing 6 per cent antimony and 1 per cent silver, have been used at low current densities and although they have strength limitations they are useful in that they can be extruded in any convenient form. Stainless steel has also been used for water tanks.

APPLICATION

GENERAL

In order that cathodic protection may be applied correctly and safely, a knowledge of all the important factors must be taken into account. These include the potential to which the metal must be lowered to prevent attack, the resistance of the water which determines the number and type of anodes which have to be used and the possibility and desirability of combining cathodic protection with other protective methods.

In an alkaline electrolyte, freshly abraded steel has a potential of -0.53 volts v. NHE (-0.85 volts v. copper-copper sulphate electrode) and, for protection, the potential of a steel or stainless-steel structure should be at least as low as this. In the presence of sulphate reducing bacteria the potential should be lowered still further to -0.63 volts v. NHE. The corresponding figures for the protection of aluminium and lead are -0.53 and -0.23 respectively.

In principle, cathodic protection can be applied whenever a metal is immersed whether in distilled water or sea water. In practice, however, the method is more readily applied in waters with low resistance and its application to waters of high resistance would rarely be economic owing to the large number of anodes necessary to get an adequate current distribution to lower the metal potential to the required value. These facts are readily understood if we consider the application of cathodic protection in estuaries and up-river where the resistance of the water may vary from 30 ohm cm for sea water to a figure of about 10,000 ohm cm. The marked effect on current requirements is readily appreciated, an anode giving 5A in sea water giving only about one-eighth of an amp in fresh water of 1000 ohm cm resistance. In a very corrosive lake which gradually increased in salinity the life of magnesium anodes gradually decreased from 4 years to one year.

In highly turbulent oxygenated water the initial current density on uncoated steel may be as high as 150 mA per sq ft. In quiescent sea water 50 mA is required initially but this reduces to 3 to 5 mA per sq ft owing to the formation of calcareous coating by the action of the current. The composition of this deposit, consisting mainly of calcium carbonate and magnesium hydroxide, varies since the amount of hydroxide increases with current density, i.e. as the pH value at the surface is increased. This deposit not only reduces the current requirements but also acts as a secondary defensive barrier if for any reason the current is switched off or reduced by a change in conditions.

In general, however, cathodic protection is used to supplement conventional paint or bituminous coatings when the current is only required to protect the metal at bare spots. Any adverse influence of the alkali formed on the metal surface on the paint coating can be avoided by control of the current, normally by fully automatic electronic devices.

The current required is less when applied in conjunction with an inhibitor and cathodic protection and chemical treatment can be made to work well together.

In applying cathodic protection to a structure it is essential to ensure that all the component parts of the surface to be protected are adequately bonded together to give a low-resistance metallic contact. Parts such as sections of piping should be properly connected by cables or metal straps.

The choice of sacrificial anodes or impressed current will depend on the size of the installation to be protected and the availability of a source of electrical power. For small systems sacrificial anodes are usually more economic. The number and distribution of the anodes will depend on the conductivity of the water and the geometry of the system. In the case of pipes an anode will only operate over several diameters even under favourable conditions hence the application of cathodic protection to pipe systems is restricted by cost to larger diameter piping.

The forms of attack which may be controlled by cathodic protection include general corrosion, pitting, graphitisation, crevice attack, stress corrosion, corrosion fatigue, cavitation erosion and bacterial attack. In controlling the pitting and crevice corrosion of 12 and 16 per cent chromium steel an adequate current leads to blistering. With nickel-copper alloys and 18/8 and 18/8/3Mo stainless steels no blistering occurs even at a current density as high as 30 mA per sq ft.

In bi-metal corrosion it is essential that the sacrificial anode should be less noble than either member of the couple, e.g. the use of zinc to protect bronze propellers and steel hulls.

The possibility of gas evolution should be considered when applying cathodic protection to prevent corrosion in confined spaces. Hydrogen begins to be evolved from steel at a potential only some 150 mV. more negative than the potential needed for protection. In these circumstances the use of magnesium anodes in sea water may not be desirable and alloyed zinc is preferred since at the low potential difference, no hydrogen is evolved. In impressed current systems the current must be carefully controlled.

At higher current densities, chlorine is evolved from sea water and chloride solutions and this is not only dangerous in confined spaces but also can attack connections above the anode. The latter can often be prevented by adequate shielding.

SHIPS

Cathodic protection is ideally suited to the protection of reserve and active ships. On a ship in motion by far the most vulnerable area is around the stern where there is intense turbulence and galvanic couples can arise between the relatively noble propeller and shaft and the steel of the hull and rudder. Elsewhere on the hull there may be paint breakdown, followed by rusting, at sites of poor adhesion such as plate joints, lines of rivets, welds, and intakes, or if there is accidental damage due to any of a variety of causes. Thus cathodic protection is often economic in reducing the high cost of docking, examination, painting and loss of sailing time.

Reserve and Idle Ships. The laying up of ships is of great concern to the navies of the world and cathodic protection is an excellent method of preservation. Not only does it permit the time between inspections to be lengthened but recommissioning can be effected much more quickly. Furthermore the application of protection is simpler than with active ships since the anodes can be disposed at any desired distance from the hull to give a more uniform current distribution. If the laid-up ships are near to a jetty, cathodic protection by the impressed current system is cheaper and the anodes may often be conveniently laid on the sea bed. In other cases portable anodes slung from the ship's side may be used to give protection during laying-up, refitting or initial fitting out.

Active Ships. In the protection of active ships the anodes are fixed along the bilge keel or around the stern or trailed behind the vessel. The whole hull may be protected or partially protected, i.e. around the stern since most of the trouble occurs in this area.

Sacrificial anodes of magnesium or zinc are usually attached to the keel and stern and must be in electrical connection with the hull. When applying impressed current the anodes must be electrically insulated and connected to the source of power through cables which must pass through the hull via water-tight glands.

There will be over-protection in the vicinity of the anode, particularly if magnesium is used, and in order to reduce this, the hull surrounding the anode is covered with a special coating. This not only avoids over-protection but has the same effect as moving the anode to a greater distance from the hull and thus ensuring a more even current distribution. Suitable coatings include vinyl resin paint, rubber or rubber-based paint, flame-sprayed polyethylene and epoxy resins. It is essential that these "patches" are applied with every possible care otherwise "holidays" may allow spread of alkali beneath the film causing loss of adhesion.

In large vessels in active service, or groups of vessels in reserve, the impressed current system is cheaper, more versatile, easier to control and damage to paint is minimised. Graphite anodes may be recessed into the hull or keel thus presenting no obstruction and thereby avoiding breakage. Other anodes which may be used are aluminium, lead alloys, platinum, platinum-clad silver and platinised titanium.

The current required will vary with changes in water temperature, conductivity, ships' speed and varying influence of the propeller. If this is not taken care of by automatic current regulation the result is either excessive deposition of scale, which increases the frictional resistance, or under-protection, which may give pitting at breaks in the protective coating. Over-protection, or the use of magnesium, can lead to the breakdown of paint if it is not resistant to alkali or has been recently applied. This sort of trouble is rare when zinc anodes are used.

The inside of bilges also presents a corrosion problem which is not easily solved. The main problem is that of any enclosed space, the possibility of hydrogen accumulation and the danger of explosion.

Trailing Anodes. For the protection of ships under way, there is some advantage in the use of impressed current with the anode trailing behind the vessel at a sufficient distance to ensure an even spread of current and adequate protection over the whole surface of the hull. A non-consumable anode of titanium or silver covered with a layer of platinum or platinum-palladium alloy may be used since both the basis metals are resistant to chlorine evolved anodically if the coating fails. Alternatively the current may be supplied to a consumable anode of aluminium wire which is automatically paid out from a drum on the stern of the ship.

There are several disadvantages to the trailing anode including troubles due to fouling, the necessity to haul the anode aboard when in a confined space, e.g. in dock, the need for skilled operation and the unsuitability of the method for use when at anchor.

Aluminium Hulls. The use of aluminium for the hulls of small ships raises the question of the suitability of cathodic protection for aluminium in sea water. In the passive condition the potential of aluminium in sea water is —0.35 to 0.45 volts (v. NHE) and it is relatively easy to lower this to the potential of —0.53 required for cathodic protection. There is a certain danger that the formation of alkali at the aluminium cathode may raise the pH value to 10 or above which is sufficient to dissolve the protective oxide film and render the aluminium active with consequent rapid corrosion. In practice, however, this is offset by the deposition of a protective calcareous layer on the aluminium surface and such troubles should not arise.

Cavitation. Pitting due to cavitation of cast-iron propellers may be prevented by the use of magnesium anodes on the tailshaft. The current density required is much greater than the 5 to 10 mA required to protect the hull and is dependent on the severity of the cavitation attack. It ranges from 30 mA per sq ft when the attack is not severe to 80 mA per sq ft when the attack is severe. The potential of the propeller must be at least 0.15 volts negative to the unprotected hull.

Cathodic protection not only reduces the corrosion components of cavitation attack but the hydrogen layer formed can reduce corrosion erosion by acting as a cushion between the collapsing vapour cavities and the surface. Good results have been obtained in reducing the pitting of propellers on launches, yachts, trawlers and super-tankers.

MARINE STRUCTURES

The protection of in-shore structures (jetties, piers, sheet-piling, sea walls, piles, dry docks, pontoons, lock and dam gates, etc.) and off-shore structures (pipelines, buoys, drilling platforms, etc.) is an important problem throughout the world. The choice of method and conditions of application will depend on the location of the installation. In-shore structures, where electrical power is available, are preferably protected by impressed current using permanent anodes. For the protection of buoys or drilling platforms, or where impressed current is impractical owing to interference with other structures, sacrificial anodes must be used. Mechanical damage to the anodes may result from ice formation or collisions with various floating objects. This may be counteracted by decreasing the size of the anodes and increasing their number or by protecting them with an insulated steel tube with a slit for the admission of water. With off-shore installations due account must be taken of the possible hazards of marine traffic, the action of waves and bottom movement.

In partially immersed structures the application of cathodic protection has advantages over the use of coatings which are very difficult and expensive to keep in repair. The part of the structure below low-tide level is continuously protected from corrosion while the tidal zone is protected

cathodically at high tide and at low tide by the calcareous layer deposited.
In the splash zone and parts above high tide where cathodic protection is
without effect, adequate protection may be achieved and maintained by the
use of normal marine paints. On other immersed parts of the structure
a protective coating should also be applied to ensure that only the minimum current density is needed. In these underwater zones it is advisable
to use resistant coatings such as asphalt and coal-tar enamels but these
are not suitable in some waters owing to attack by teredos or marine
borers. In such cases sheathing with monel, plastics, e.g. polyvinylchloride or polyethylene, or concrete may be used. An epoxy resin coating
is also advantageous, particularly in the tidal zone, where it can be applied
as a thick viscous quick-setting layer.

If the footings of a marine structure are buried in mud in which sulphate-
reducing bacteria are likely to proliferate, the potential of the structure
must be depressed to a lower value in the range —0.63 to 0.68 volts (NHE).

On submerged pipelines while it is advantageous to apply a supplementary
coating to that applied by the manufacturer this is often unnecessary since
the current required will be soon reduced by the formation of a protective
calcareous coating.

STORAGE AND BALLAST TANKS

Cathodic protection may also be used for the protection of a variety of
tanks in which sea water is stored or used as ballast. It is common practice for tankers to make one journey full of oil and the return journey with
the tanks full of sea water. In other cases tanks may be alternately filled
with oil (or oil products) and sea water, as in refineries where sour oil
may be pumped from tank to ship by displacement with sea water. Although on first consideration either impressed current or sacrificial
methods might appear to be appropriate the application of cathodic protection is here restricted by spark hazards. Furthermore precautions must
be taken to ensure that the current is not so high as to cause the liberation
of chlorine which could attack the structure above the waterline and constitute a danger to personnel. The use of sacrificial anodes is often preferred to impressed current because with any electrical installation there
is always a possible fire risk, however remote this may be. There is also
a spark hazard attached to the use of magnesium hence the heavier zinc
is usually favoured although its lower driving potential necessitates the
use of a greater number of anodes. The efficiency of this form of protection is adversely affected by the oil film formed on the surface of the
anode.

In refineries using sea water for cooling the protection of large storage
tanks and box-coolers by cathodic protection can provide a significant
saving since protective linings may be dispensed with entirely.

The use of cathodic protection of tanks is only useful while the tanks are
in ballast and other methods of protection must be sought when the tanks
are empty and their walls are wet and exposed to the air. To overcome
corrosion during this period effort has been directly to the incorporation

of inhibitors in the water or oil before the tank is emptied. This has met with varying success but much research still remains to be done. An oil tank is not the simple structure its description would imply and a number of possible sites for attack are built into it. The steel is usually partially covered with mill scale, there are numerous crevices at junctions of girders and pipes, sedimentation can occur and with the heavier oils it is necessary to install a heating coil which may be of a non-ferrous metal.

CONDENSERS AND HEAT EXCHANGERS

In the past it was common practice to use Admiralty-brass condenser tubes and ferrules in naval brass end-plates with ferrous water boxes. The copper alloys were cathodically protected by the iron or steel which underwent enhanced attack necessitating a corresponding increase in thickness. Alternatively zinc protectors could be used to protect both the ferrous and copper alloys. This protection also extended into the condenser tubes for a distance of about 2.5 diameters and reduced impingement attack which was most severe there.

If the water boxes are coated or replaced by reinforced plastics, and if the doors are made of gunmetal, steps must be taken to restrict the attack on the end plates and the tube ends. It was at one time standard practice to use iron protector slabs which acted as sacrificial anodes and in addition the iron corrosion products contributed to the formation of a protective film on the condenser tubes.

The replacement of Admiralty brass by aluminium brass and cupronickels has eliminated most of the corrosion troubles. Those that do occasionally occur are due to the accumulation of debris which causes local turbulence, impingement and deposit attack. Where cast iron or steel doors are used corrosion has been successfully prevented by the use of magnesium anodes or impressed current. In coastal power stations cathodic protection has also been employed for water boxes of condensers which use sea water for cooling. Titanium anodes with automatic current control are particularly suitable for the task.

FRESH-WATER TANKS

The cathodic protection of hot and cold water tanks by the use of magnesium anodes depends on the total dissolved solids being in the range 40 to 400 ppm. About 80 per cent of natural waters fall within these limits. If the waters are very soft and contain less than about 40 ppm dissolved solids (about 5 per cent of waters) the resistance is too high for adequate protection to be provided. Very hard waters containing more than 400 ppm dissolved solids (13 per cent waters) give rise to gas evolution with magnesium anodes which is not only objectionable but may be hazardous. In these cases impressed current or, in the case of cold water tanks, zinc, may be used.

The protection of galvanised steel hot water tanks by magnesium anodes can reduce the loss of zinc by a factor of 5 to 30. A magnesium anode is simple to fit and for rectangular tanks may be a 2 lb. disc attached to the inspection plate. For cylinders, a vertical rod anode may be attached to the boss at the top of the dome. With indirect cylinders one or two rod anodes must be placed in the space between the calorifier and its outer wall, the number and spacing of the anodes depending on the design.

Steel tanks may be protected under a wider range of conditions than galvanised tanks owing to the absence of the zinc-iron potential reversal at elevated temperatures. This means, of course, that zinc cannot be used as a protective anode in hot-water tanks.

INTERNAL PROTECTION OF WATER PIPELINES

For the internal protection of large diameter pipelines it is usually adequate to use the ordinary coatings. However, with the greater use of thinner gauge steel rather than cast iron, the prevention of corrosion is more important and in acid soft waters or waters of high salinity it is possible to use cathodic protection. This must, of course, be restricted to the larger pipes and, even then, there are many difficulties to overcome. Nevertheless there are many instances when it is not only practicable but also economically justified. Sacrificial anodes cannot be used because parts can become detached after partial dissolution and cause blockages at valves etc. downstream. Furthermore the pipelines or anode may be relatively inaccessible which makes replacement difficult and expensive. Impressed current must be used with permanent anodes. These anodes must be placed centrally in the pipe and be small so as not to restrict the water flow. Special arrangement has to be made so that they may hang, but be insulated, from the pipe wall (Fig. 41). This can be effected by welding to the pipe wall a flanged neck containing an insulating bush, e.g. neoprene, through which is passed the anode support rod to which the electrical con-

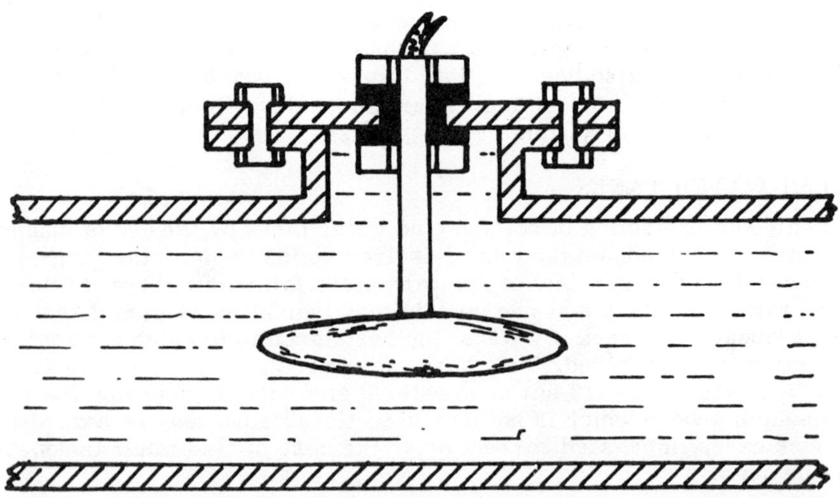

Fig. 41.   METHOD OF ATTACHMENT OF INTERNAL ANODE
FOR THE CATHODIC PROTECTION OF PIPE LINE

nection is made. Sealing is completed by a suitable flange plate. Lead anodes are suitable and are cheap but users may favour ultimately the platinised anodes to which reference has already been made. The dispositions of the anodes along the pipe will depend on the resistivity of the water and the diameter of the pipe. The pipe should be coated initially to minimise the current needed and suitable precautions must be taken to avoid overload and possible stripping of the coating.

ANODIC PROTECTION

Although it is unlikely that anodic protection will be used in natural waters it has applications in other fields. When using cathodic protection an attempt is made to ensure immunity from corrosion, but many methods of protection depend for success on the maintenance of passivity, e.g. the use of inhibitors and high pH value. Anodic protection is in this second, passivity class and experimental work has shown that it is not only possible but that it may have definite applications in chemical plant where other methods may be impracticable. There are certain limitations in the process, not least being the fact that to ensure complete safety the potential must be maintained between strictly controlled limits. This means auxiliary monitoring equipment for controlling automatically the current and potential.

Chapter XII

DESIGN

At all stages of the design, construction and operation of plant, adequate attention must be given to the possibility of corrosion otherwise there may be costly failures and hold up of production. All too often the interested parties lack sufficient experience to foresee corrosion troubles and the corrosion consultant is called in when plant failures have already occurred. There is no need for this providing the designer, manufacturer and user are conversant with the factors producing corrosion or seek the advice of experts. Corrosion prevention should start on the drawing board and adequate consultation should take place frequently during the development of the project.

An over-riding factor will usually be one of cost. This will not only include the capital cost of the equipment but also the cost of maintenance, replacement and possible shut-down. The contractor will want to keep his estimate competitive but this may involve the use of less resistant materials. Consequently it may not be cheaper in the long run for the buyer to accept the lowest tender.

The distinguishing characteristics of a good design include adequate protection against corrosion and provision for ease of inspection and maintenance. Constructional materials should be chosen not only to meet the engineering requirements but also for their resistance and ease of protection against corrosion. If the equipment is to be used with waters of different or varying composition, due allowance will have to be made for the variation in corrosivity and the availability of skilled personnel necessary to maintain control.

In the choice of materials, direct service experience is the best guide and if any testing of materials is carried out, the experimental conditions should be as close as possible to those which will eventually apply. Tables are available giving the corrosion rates and behaviour in various media but they can act only as a rough guide for materials under service conditions.

Corrosion can seriously impair mechanical properties and a sufficient margin of safety must be allowed to take care of such things as reduction of cross-section. In the cases of localised attack these measures may not be adequate. The possible influence of the environment on stress and fatigue failure should never be forgotten, both the permissible stress and the fatigue limits being lowered when corrosion occurs.

In many cases it may be necessary to employ surface coatings with greater corrosion resistance than the load-bearing metal or to separate the metal from its environment. In other cases the corrosivity of the environment may have to be reduced by the addition of inhibitors or water treatment, and even the process itself may have to be changed so that the conditions are less severe. Alternatively, cathodic protection may be applied but if use is made of impressed current any possible damage due to electrical failure must be considered.

CHOICE OF METAL

Metals are selected primarily on the basis of mechanical and physical properties such as tensile strength, ductility, and resistance to mechanical damage and thermal shock. Other important factors are the ease of fabrication and repair, which may involve welding, soldering and brazing. The cost of the basis material and of fabrication must be considered as well as whether the particular alloy, in the required form, is available. Stress should also be placed on the corrosion resistance which not only appreciably affects the cost of maintenance and determines the useful life of the installation but also may affect the purity of the manufactured product. Important factors are the susceptibility of the metal either to localised attack leading to leakage or severe general attack which weakens sections so that they are unable to withstand the stresses imposed and fail prematurely. Use of the more resistant metals is limited by cost and corrosive conditions must be really severe before the use of rare metals and expensive alloys can be justified.

The most convenient and economic metals are steel, iron, galvanised steel, lead, copper and aluminium with brasses widely used for fittings and stainless steels for special purposes. The use of plastics is increasing year by year as additional information on their behaviour overcomes any doubts of the more conservative users.

The poor corrosion resistance of iron and steel limits their use unprotected to cases where corrosion may be allowed for by increased section. They are still pre-eminent as structural materials in boilers. Small additions of copper, up to 0.25 per cent, are beneficial under slightly acid conditions and if other requirements permit, designers should seriously consider the use of copper-bearing steels where there is any likelihood of low pH values developing.

Galvanised steel has superior corrosion behaviour and is satisfactory in most waters. It is not suitable for soft waters or waters with a high free carbon dioxide content. The corrosion resistance is enhanced by the formation of protective scales over the metal surface. Those formed from river waters are often more useful in this respect than those from well waters. Pitting failures are often due to deposition of debris, e.g. iron filings or inert material, which interferes with the formation of a protective scale. To reduce the possibility of this type of failure tanks should always be covered.

Lead is very flexible and can be fabricated easily but it has low strength and is liable to failure by creep or fatigue. To resist mechanical damage and bursting, pipes must be heavy-walled and horizontal runs must be adequately supported. Lead is attacked by certain waters with attendant danger to health since lead is a cumulative poison.

Copper is widely used for plumbing and in the half-hard condition can be bent cold and being rigid and light requires only a few supports. Piping for domestic water systems can be joined by capillary soldering or compression fittings, while on a larger scale, neat and strong joints in light gauge copper can be made by autogenous welding or bronze welding.

There is a possible danger in the use of copper pipes with galvanised hot-water tanks since pitting of the latter may be stimulated by copper concentrations greater than about 0.1 per cent. Copper-bearing water from the hot tank can return to the cold tank if the tanks are positioned too close together or, via the expansion pipe, if the water is allowed to boil.

Aluminium is very resistant to pure and very soft waters and, if the pitting tendency were reduced, could become a useful and desirable replacement for galvanised iron, copper or lead which are not entirely satisfactory. The pitting may be reduced by cladding with a more anodic alloy, e.g. one containing one per cent zinc.

Aluminium is more sensitive than zinc to dissolved copper. Water discharged from a copper system may contain traces of copper and it is incumbent on the contractor to warn the user of these dangers when the water is cupro-solvent.

Light alloys are widely used in marine construction. Both pure aluminium and its alloys have a high resistance to corrosion by sea water and both the aluminium-manganese and aluminium-copper alloys have reasonably high tensile strengths.

Brass, used in many fittings, gives no trouble except for dezincification in soft, acid or brackish waters. If the attack is of the 'plug' type seepage of water occurs. In dezincification the dissolved zinc may form voluminous corrosion products which block the waterways. Owing to the deterioration in strength, unions may break when unscrewed and seats on taps or valves be readily eroded by water.

The more widespread use of stainless steel is prevented by both cost and susceptibility to attack in waters containing moderate amounts of chlorides. They are useful, however, where high strength and low cross sections are required, particularly in high temperature environments.

Various rigid forms of plastics may be used for cold water tanks. Fibre-bonded plastics find use as bushes and bearings in sea water and are very satisfactory as insulating materials. Glass fibre materials are light and strong and free from corrosion troubles and may be used for small boats and tanks. A wide variety of plastic materials is finding application in water distribution networks and waste disposal systems. In many cases, however, experience with such materials has been of too short a duration to enable the long-term behaviour to be predicted.

The possibility of bimetal corrosion, i.e. galvanic couples resulting from contact of different metals, should always be borne in mind. In equipment for use in hard or moderately hard waters the dangers are often negligible but in equipment that is to be used in all kinds of waters potentially dangerous bimetal couples are better avoided.

Since the effect is often proportional to the ratio of the area of the cathode to the area of the anode, combinations should be avoided where the area of the cathodic metal is relatively large. The reverse case is tolerated and may be an advantage as in the joining of mild steel flanged joints with more resistant alloy bolts or in riveting or welding of mild steel with a slightly more noble low-alloy steel.

Troubles due to galvanic couples may be eliminated by the use of electrical insulation of the two metals, coatings or jointing materials, inhibitors or cathodic protection. Insulating washers are often used (Fig. 42) and also non-hardening jointing compound around the area where the two metals are in contact. The electrical insulation of copper in cupro-solvent water is not enough and the copper dissolution should be prevented by nickel plating or some other coating.

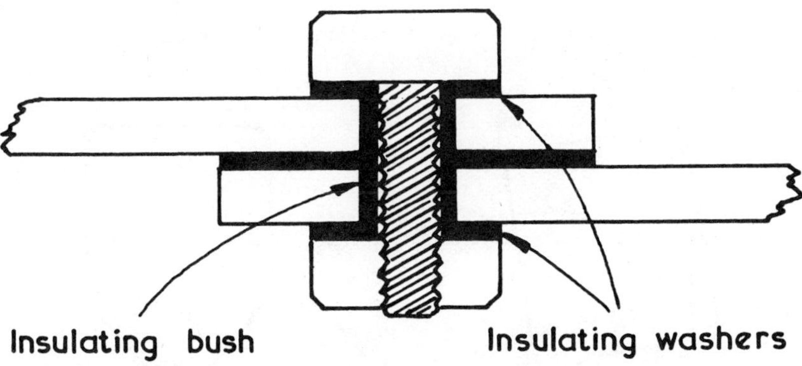

Fig. 42. METHOD OF ELECTRICALLY INSULATING DISSIMILAR METALS

An alternative method is to apply a zinc chromate primer to all the mating surfaces and to paint the region on either side of the joint. If the whole area of one metal is to be coated with paint it should be the cathodic and not the anodic member of the couple since coating the latter would result in intensification of attack at any defect or bare spot.

If dissimilar metal couples have to be used it is better to use brazed or welded joints rather than threaded joints. If the latter are unavoidable, say, in piping systems, the first section of the less noble metal should have a heavier wall at the joint and preferably it should be a short section located so that it can be replaced readily. In joining plates, the thickness of the less noble metal may be increased in the vicinity of the joint.

GEOMETRY AND WORKING CONDITIONS

Even in the simplest of designs, the desire for cheapness and ease of
fabrication may prevail over common sense. For example, the types of
container shown at top left in Fig. 43 are far too common though it is
obvious that they can never be drained. There is very little cost involved
in producing the better designs shown. Most cases are, of course, not so
easily solved and the factors involved are usually varied and complicated.

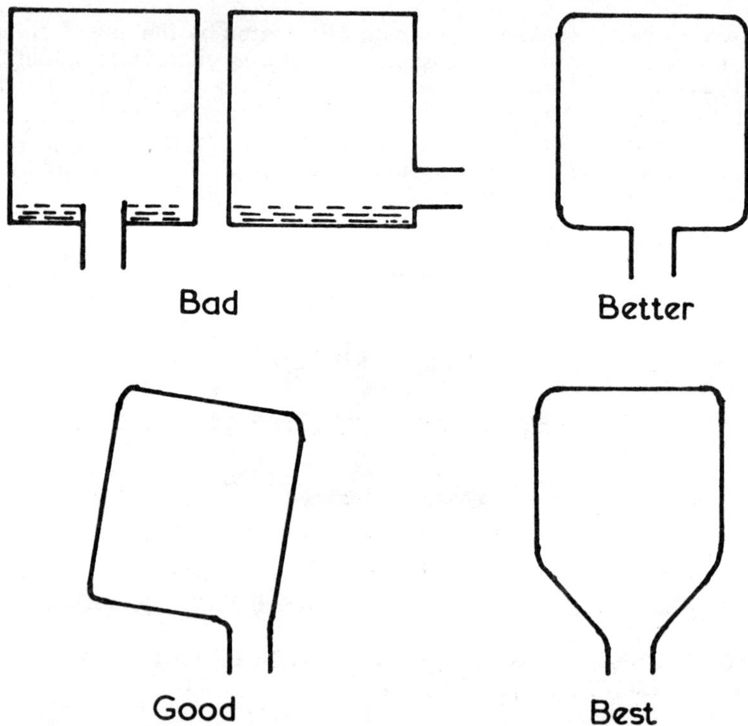

Fig. 43.   POOR AND GOOD DESIGN OF A SIMPLE
CONTAINER

MOVEMENT

In order to avoid corrosion, corrosion-erosion and cavitation, attention
should be given to the velocity of the water. As we have already discussed,
increasing velocity generally increases the corrosion of metals. A certain
degree of movement is often desirable to make the environment more
uniform and reduce the likelihood of localised attack. This is especially
true of metals such as aluminium and stainless steel. In stagnant sea

water austenitic stainless steels are liable to pitting, but as pump impellers in the same water, they give excellent performance.

Biological fouling is also more likely at low velocities but above a certain flow speed the ability of such organisms to adhere is appreciably impaired, e.g. below 2 ft/sec barnacle growth is possible and deep pitting may occur.

At high velocities, on the other hand, suspended solids or gas bubles can give rise to corrosion-erosion and impingement, while stiller higher velocities produce cavitation. A small amount of general turbulence only leads to serious attack when resistance to corrosion is inherently low. Localised turbulence on the other hand should be avoided as far as possible since differences in velocity can give rise to galvanic cells. Such promoters of swirling and turbulence are flanged joints; sharp changes in cross section, e.g. at valves; nozzles and other constrictions; sharp bends and any obstacles to flow.

Alloys which depend for their corrosion resistance on the formation of thin protective films actually benefit from a moderate degree of turbulence. On the other hand alloys which are protected by a relatively gross film of corrosion product, that may not be retained under turbulent conditions, may suffer serious damage if the velocity is too high.

High or low velocities should thus be avoided and a suitable general range of water velocity is 2 to 8 ft/sec but this is very dependent on the metal and the water composition. Thus a marked difference exists between the behaviour of brass and stainless steel in brackish water. Brass tubes should not be used above a velocity of 5 ft/sec while the minimum velocity for stainless steel is 5 ft/sec.

The best materials for use in marine condensers and cooling systems with water at high flow velocities are aluminium brass and cupro-nickel alloys. The cheaper aluminium-brass is more widely used but where durability is more important than the expense, e.g. in naval use, the more costly 70/30 copper-nickel alloy is employed. For high-pressure sea water supplies, the 94/5/1 copper-nickel-iron alloy is found to be more resistant than copper but even with this alloy local turbulence should be kept to an absolute minimum.

In the choice of pipe diameter for high velocity water flow a balance must be made between the cost of the power used in driving the fluid through a smaller pipe and the cost of installing larger pipes. This has lead to the concept of an 'economic velocity' which is usually a few ft per sec.

Both impingement and cavitation damage can be reduced by keeping the absolute pressure as high as possible to restrict the release of gas bubbles. This can be done by controlling the flow at the outlet rather than at the inlet to the system and by free discharge of the water rather than discharge into a long pipe which can reduce the pressure by a syphoning action.

The water velocity in nozzles or other constrictions should be kept low and the selection of valve design should take into account the influence of change in velocity in accelerating galvanic attack. The butterfly type of valve has the advantage over most others in that in the open position it offers the minimum resistance to flow. It may be rubber-coated if desired and is only 60 to 80 per cent of the weight of the normal type of valve.

ELEVATED TEMPERATURES AND HEAT TRANSFER

In condensers and radiators the presence of scale, corrosion product or dirt not only reduces the thermal efficiency but also the water flow speed. A more resistant material may be an economic alternative since a longer life may be obtained from the same equipment or, what is more important, the size of the system and the flow of water may be reduced.

In heat exchangers some allowance can be made for corrosion but this is only possible on some parts, e.g. 250 mils may be added to the thickness of the shell and tube plate but the tubes must be made of more resistant and expensive material. In heat transfer, account must be taken of the difference in temperature of the bulk of the water and that of the metal surface which may be raised considerably for a given heat flux. The surface temperature may be raised still higher in the presence of scale and severe corrosion can result. In boilers, deposits on the tubes may raise the temperature to a level at which the tubes oxidise, distort, and finally burst. At lower temperatures, as in the water spaces of an internal combustion engine, excessive scale may mean loss of cooling efficiency and overheating.

Since gas and vapour have low thermal conductivities any stationary bubbles or gas vapour trapped on a metal surface will reduce the transfer of heat and increased corrosion will take place owing to the rise in temperature. This may be minimised by tilting any such surface at a slight angle and ensuring the escape of any steam formed.

In heat transfer equipment the peak metal temperature can be reduced by ensuring that the coolant and the liquid or vapour being cooled flow in the same and not the opposite directions.

Severe corrosion of mild steel in condensate containing small amounts of oxygen and carbon dioxide may be reduced, with some slight loss of efficiency, by continuously venting the steam so that the concentration of the gas is lowered.

For scale-forming waters the corrosion may be reduced by water treatment, the use of a small closed cooling system, indirect cooling by the main cooling water supply, or the 'cooler' may with advantage be replaced by a waste-heat boiler with properly conditioned water.

In design, if adequate provision is not made for filtering dirt and silt from in-shore waters used for cooling, excessive corrosion can occur. 70/30 copper-nickel alloy is susceptible to deposit attack and there may be rapid pitting of dirty tubes at high rates of heat transfer.

Both copper and galvanised steel are suitable for hot-water storage tanks. By making the tanks cylindrical with domed tops and bottoms a strong tank can be made from light gauge copper sheet. Galvanised tanks may

be cylindrical or rectangular and are usually used with galvanised piping
and cast iron boilers. Corrosion can be caused by persistent over-heating
and the consequent reversal of potential between zinc and steel. The life
of a tank can be increased by years by lowering the temperature by 5 to
10°C and in this respect thermostatted systems are to be recommended.
Intermittent operation of large capacity immersion heaters is preferable
to continuous operation of heaters of small capacity.

JOINTS

Soldering and brazing. Copper, brass, iron, zinc and galvanised iron, and
tin-plate are frequently joined by soldering or brazing. It is essential
that the metal be clean and wetted by the solder. The thin oxide film
present is removed by use of a flux, usually zinc or ammonium chloride,
and care should be taken to wash the joint thoroughly afterwards to get
rid of any remaining chloride which would otherwise cause increased
attack. Seams in copper are usually brazed with brazing brasses containing 40-50 per cent zinc which give good service but are susceptible
to dezincification in acid and brackish waters. This may be avoided by
brazing with an 81/14/5 copper-silver phosphorus alloy which is more
expensive but is fairly ductile and with capillary seams makes an economical and good job.

One of the advantages of lead is that soldered joints and tees can be made
by wiping with no necessity for special fittings. This procedure is made
possible by the use of plumbers' solder, 33/67 tin-lead alloy, which is
plastic in the range 183 to 250°C allowing sufficient time for the joints in
lead pipes to be wiped and shaped.

Riveting. In the use of rivets care should be taken to limit the pitch and
to ensure that the holes and, in the use of countersinking, the depressions
are completely filled. The rivets can with advantage be made from a
slightly more noble and resistant metal, e.g. low alloy steel for riveting
mild steel. The main disadvantage of this method of assembly is the
crevice which can lead to enhanced corrosion in a number of ways.

Screwed joints. Screwed jointing is used for demountable connections
and invariably for joining mild steel or galvanised iron piping. This has
the disadvantage that a full range of bends, tees and other junctions must
be available and the process of cutting tubes to length, threading and
screwing up the joints is slow. Where long runs of piping are involved
the use of galvanised steel is economic but for complicated runs it may
be more expensive than copper or lead because of the higher installation
costs.

Joints like the one shown in Fig 44 should be avoided since enhanced
attack can occur not only owing to the additional turbulence produced by
the sharp variation in cross section but also by failure to remove the burr
on cutting the pipe. Furthermore in water of low conductivity the cathodic
protection afforded by galvanising will not be effective over any great
distance.

Fig. 44.   ADDITIONAL TURBULENCE AT SCREWED JOINT

Flanged joints. Troubles, e.g. impingement attack, can arise with flanged joints owing to slightly misaligned flanges or to gaskets projecting into the bore (Fig. 45). It is very difficult to cut a washer to an accurate enough dimension so that a smooth bore will be produced when the flange is tightened and the gasket is compressed. This difficulty is more satisfactorily solved by the use of spigotted flanges or by utilising 0-rings which may be of rubber, metal or gas-filled metal.

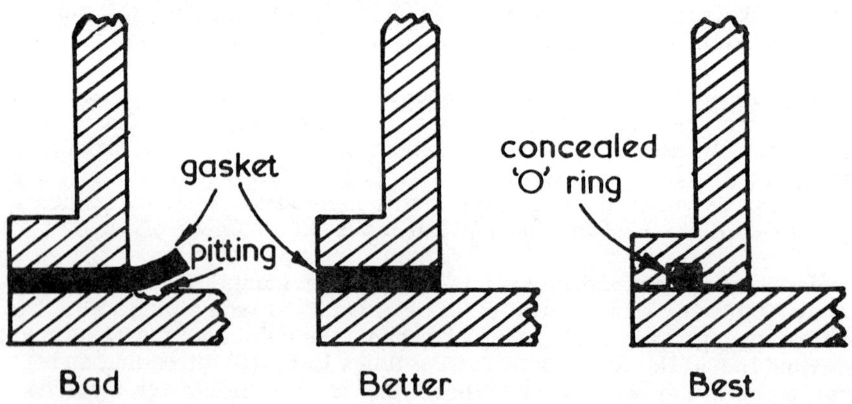

Fig. 45.   DESIGN OF FLANGED JOINTS

In flanged stainless-steel joints the use of fibrous gaskets should be avoided since by capillary or wick action the fibres draw solution into the joint and localised attack is promoted on the flange faces. Polyethylene, Teflon (PTFE) or neoprene do not have these disadvantages.

Welds. Butt-welded joints and continuous seam welding are preferred to riveted or bolted lap joints and, if the latter are used, it is advisable to fill or weld both sides of the crevice in order to avoid access and retention of liquid. Care must be taken in the choice of welding rod so that the weld metal does not differ substantially from the base metal. The former can with advantage be slightly more noble to ensure that the weld region is slightly more cathodic to the rest of the surface and no preferential attack takes place.

The preparation, welding technique and finishing should be carefully carried out so that not only is the final weld relatively smooth and well-shaped, but also does not create areas susceptible to local attack because of the presence of porosity, voids, crevices, scale inclusions or changes in metal structure. This is aggravated by excessive heating such as occurs when two pieces of metal of different thickness are welded together. Heat is not dissipated as rapidly through the thinner metal section which consequently reaches a higher temperature thus creating a variation in the grain size and structure. The thicker section component should be machined down so that the thickness is uniform. If possible the machined part should not be put in contact with water since it may be preferentially attacked.

The most common materials fabricated in this way are mild steel, stainless steel, copper and aluminium alloys. With stainless steel there is a danger of intergranular attack since, owing to the temperature gradient, some of the parent metal will be in the temperature range 400 to 900°C in which carbide precipitation may occur at grain boundaries. Failure is likely to occur unless the weld is heat treated subsequently. The welding rod should be chosen with care, the basis metal should preferably be a stabilised low-carbon austenitic chromium-nickel steel and where possible the weld should be finally surface-ground.

STRESS

In specifying constructional procedures, care should be taken to minimise conditions likely to lead to stress corrosion. The highest stresses are likely to be set up by welding and adequate stress-relieving treatment should be specified. Limitations should be placed on the degree to which parts are 'force-fitted' in order to avoid concentration of stress while, in parts subject to vibrations, stresses should be reduced to a minimum to lessen the chance of failure by corrosion fatigue. Account should be taken of the increase in working stress and reduction of the fatigue limit produced by a reduction in cross-section due to corrosion.

It is usually extremely difficult to heat-treat large structures on site. Hence it is essential to design so that stress relief on individual parts is possible prior to erection. Failing this, materials and methods of fabrication must be chosen so that satisfactory resistance to corrosion and a low stress level are achieved. High-chromium ferritic alloys such as 20/12/3 chromium-nickel-molybdenum or 25/20 chromium-nickel are more resistant to stress corrosion than austenitic steel.

When stress relieving is specified it should be remembered that riveted joints and expansion fits will tend to loosen on heat treatment and, therefore, it is desirable to use welded joints throughout.

CORNERS AND EDGES

Sharp edges, corners and recesses should be avoided and wherever possible all contours should be rounded and given as large a radius as possible (Fig. 46). This makes the installation easier to clean, metal coat or paint, thus ensuring better continuity, adhesion and uniform thickness of coating, and better resistance to corrosion.

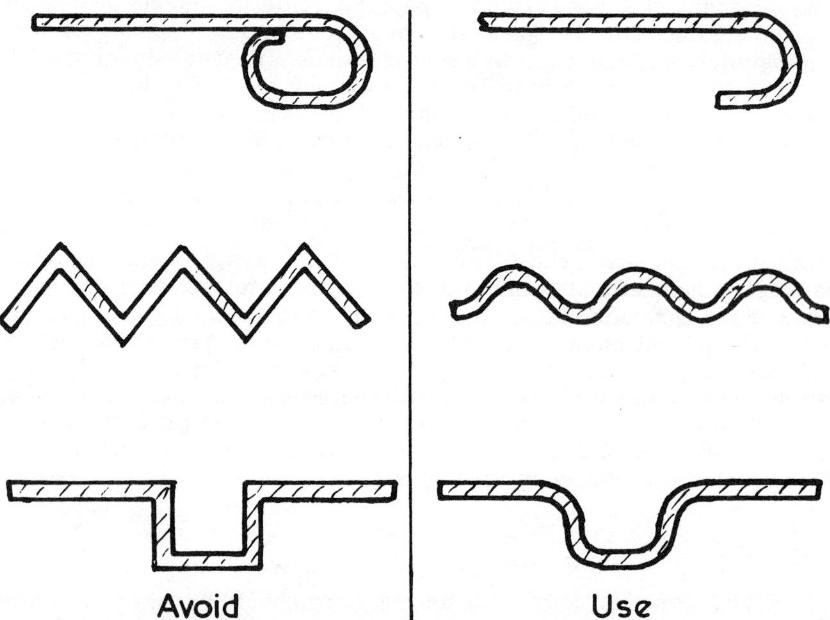

Fig. 46.   ELIMINATION OF CORROSION SITES AT SHARP CORNERS AND RECESSES BY SIMPLE CHANGE OF FORM

A complex shape will lead to high plating costs since the total weight over the whole article will have to be greater in order to ensure that the thickness is adequate over parts which are difficult to plate (Fig. 47). In metal spraying it is very difficult to ensure that the thickness in corners is adequate owing to the poor accessibility.

Rapid change of contour, the heads of rivets and screws, and crevices in which pickling and pretreatment solutions may be trapped, provide regions where subsequent paint coating is likely to breakdown.

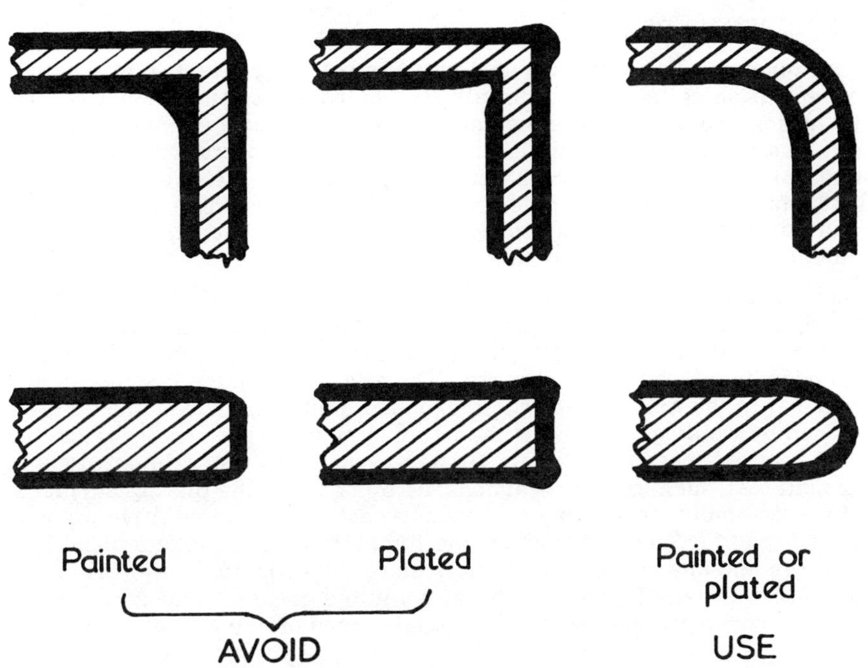

Fig. 47. VARIATION OF COATING THICKNESS ON CORNERS AND EDGES

On a smaller dimensional scale, the roughness of the surface should be sufficient to provide adhesion for paint or sprayed metal but not too great that even coverage is affected (Fig. 48). The primer of first coat should be able to cover the surface adequately. A useful guide is that the surface roughness should not be greater than about one-third of the coating thickness. Since a desirable coating thickness is 3 to 5 mils the surface roughness should not exceed about one mil and deep scratches and imperfections should be removed by sanding or grinding.

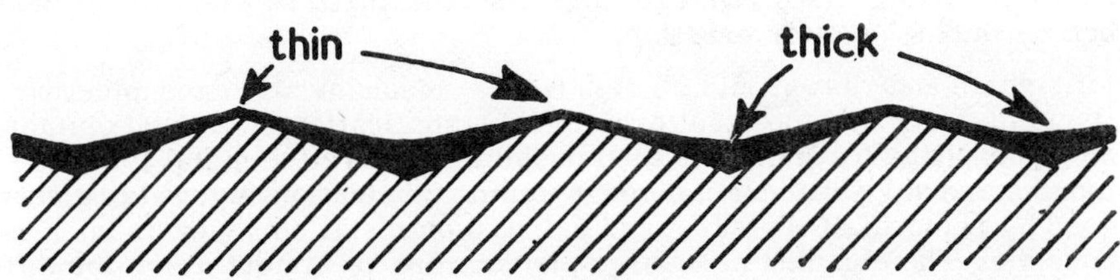

Fig. 48 EFFECT OF ROUGHNESS ON THE VARIATION OF THE THICKNESS OF PAINT COATING

CREVICES AND DEPOSITS

It is impossible to avoid crevices entirely but their effect in producing
concentration cells can be minimised. The nature of the contacting material, which may be metal or nonmetal, can have a marked effect. Although
stainless steel would normally be compatible with bronze, a stainless
steel spindle in a bronze bush may undergo intense attack near the mating surfaces. Rubber and plastics may, for reasons which are not clear,
excite differing degrees of attack. In the case of rubber, the free sulphur
content is likely to be important.

Deposits of loose rust or scale, cotton waste and metallic particles or
turnings all form crevices and can and should be avoided. Larger crevices
are formed at all flanged or other pipe joints, riveted or bolted seams,
expanded-in tubes and fitted parts. The attack within the crevice is proportional to the area of the freely exposed metal. Crevices should be
avoided by welding, improving the fit or by making the crevice wider than
about 5 mils. Ingress of water can be prevented by using impervious jointing material, such as fluorocarbons, or by painting the mating surfaces
before assembly, preferably with paints containing an inhibitive pigment,
i.e. red lead for steel, zinc chromate for aluminium. Gaskets should not be
too large and any caulking materials should not form harmful crevices at
their outer edges. In many joints an inhibited grease composed of lanolin
and zinc chromate may with advantage be used on the gasket.

PROTECTIVE COATINGS

If it is decided that a coating will be required to increase the corrosion
resistance, the whole design must be worked out with the coating in mind.
The use of a more resistant metal as a coating can only be justified if it
is more economic than the use of the solid metal. Factors to be taken
into account are whether the appropriate coating can be applied and not
damaged during assembly of the equipment, the probable life of the coating
and whether it can be renewed *in situ*. Consideration must also be given
to the resistance of the coating to abrasion and impact, its thermal stability and its effect on heat transfer.

Most coatings applied to iron or steel cause local corrosion if they are
inadequate. This may be inherent or created by poor pretreatment, application or maintenance. Defects may also be caused by mechanical damage
before, during or after erection.

Mill-scale and rust should be removed by mechanical means, pickling or
flame cleaning and the steel protected by application of an etch primer
or a phosphate coating. It is then possible to store the metal for some
months and any subsequent damage during fabrication, e.g. welding, made
good by removal of the scale and painting or metal spraying as soon as
possible. The cleaning processes and priming treatment differ for different metals hence it is not easy to clean articles made from a number of
metals. In such a case it may be preferable to coat the different metals
separately before assembly.

The vacuum-blasting technique enables mechanical cleaning to be carried out in a confined space without danger to the operator.

Aluminium or zinc coatings with a thickness of 3 to 5 mils not only give excellent protection in themselves but provide a good key for paints. Under conditions of full immersion aluminium-sprayed coatings are better and cheaper than zinc and have remained intact in sea water for 9 years. In a marine atmosphere both are good but zinc is slightly superior.

Although these coatings have good corrosion resistance it is advisable to apply paint wherever possible. Such a system has considerable long-term economic advantages over paint systems alone. It should be borne in mind that antifouling paints containing copper or mercury should be avoided on an aluminium structure.

The use of plastics coatings is increasing and pipe-lines may, for example, be coated internally with baked-on phenol-formaldehyde resins, and in conditions involving high pressures and velocities at low temperature, steel pipework may be dip-coated with PVC. Oil refineries have used cement linings combined with a neoprene coating. The life of bronze propellers and pump impellers can in some cases be doubled by a neoprene coating which is one of the few coatings that can temporarily alleviate damage caused by cavitation-erosion or impingement attack. The coating is very firmly bonded to freshly shot-blasted and primed surfaces.

Epoxy paints based on pitch give good protection to the internal surfaces of oil tankers as have epoxy resin polyamide paints on the hulls of passenger liners.

CATHODIC PROTECTION

Cathodic protection is used for shore installations, e.g. piers and jetties, the hulls of ships in reserve or in active service, heat-exchange equipment and tanks and reservoirs (Chap. XI.). The primary requirement is that the water should have a reasonably high conductivity hence it finds its widest application in installations using sea water and brackish water. It is usually used in conjunction with paints thereby reducing the current requirement and preventing corrosion at any breakdown in the coating.

Magnesium anodes may be used to protect galvanised hot and cold water tanks, the zinc, the alloy layer and the steel all being protected. By the time the anode has been consumed sufficient scale will generally have been formed on the metal to make renewal unnecessary. Cathodic protection is not usually applicable to soft waters because the low conductivity makes it difficult to maintain an adequate current distribution.

The cathodic protection of the interior of condenser water boxes is established practice on some shipping lines and shore installations and could be universally adopted. It prevents the graphitisation of cast iron, dezincification of gun-metal doors and attack at the end of tubes. Ships' hulls may be protected by fixed or trailing anodes which may be either sacrificial or permanent with impressed current.

The successful application of resistant alloys often relies on the cathodic protection afforded by other less-noble metals.

The corrosion of cupro-nickel alloys, e.g. K-monel, 66/29/3/2 nickel-copper-aluminium-iron, used for propeller shafts on high speed craft, is reduced considerably by scale deposited in the cathodic protection given by the anodic hull, hull fittings and propeller. Cathodic protection can fail if there is any shielding of the surface, e.g. fouling, and severe pitting may result. The cathodic scale deposited may in some cases be an inconvenience, e.g. on moving parts with close tolerances. These conditions can be obtained by the chance proximity of dissimilar metals but if these are insulated, appreciable attack of the previously protected metal may be caused.

INHIBITORS

The criteria governing the possibility of the use and choice of inhibitor are many and various (Chap. IX). With any inhibitor which can give trouble if a critical concentration is not maintained, a governing factor will be the availability of personnel who are able to make regular checks on the inhibitor concentration. In large systems and in those in which the water is to be used for some processing, e.g. of comestibles, the choice is limited both by cost and toxicity. To the initial cost of the inhibitor should be added the cost of maintenance and chemical supervision.

One must consider the possible influence of the inhibitor on the production of scale and sludge, formation of surface coatings affecting heat transfer and the loosening of any existing scale leading to blockage of lines. The production of foaming by some inhibitors and the problem of disposal are also limiting factors.

CONSTRUCTION

The specification should be sufficiently comprehensive to include not only the constructional methods but also precautions to be taken to minimise corrosion troubles when the plant comes into operation. Even the smallest detail should not be overlooked, e.g. the carelessness of leaving iron filings or turnings in a galvanised tank can initiate serious local attack. The manufacturer or contractor should be required to provide guarantees that all recommended precautions have been carried out.

A very important improvement could often be made in the storage of components prior to erection. All too frequently parts are left lying about a site for months at the mercy of corrosion by the atmosphere. While this may not matter for components that are to be mechanically descaled after erection, it is a most undesirable procedure with items such as piping and radiators in which the internal corroded surface cannot be easily restored to an acceptable condition. Rust and dirt can cause damage to valves and turbine blades, block small orifices and generally interfere not only with the operation of the plant—and possibly contaminate the product— but also with any corrosion protection measures taken subsequently.

The cost of keeping hollow steel components clean and free from corrosion by plugging outlets and keeping the interior dry by the insertion of a little quicklime during transport and storage can show a high return in later stages.

Pickled and primed steel plate should be handled on site with sufficient care to minimise damage to the priming paint and any damage suffered in fitting should be made good as soon as possible.

The design should not only include specifications of the materials to be used but also instructions for inspection of materials and fabrication methods at all stages. The precautions to be taken in welding and heat treatment after welding should be included and the design should be such that these operations can be readily carried out.

Care should be taken when specifying paints and painting procedure that surface preparation is included.

Inspection of the equipment should not exclude the use of X ray, γ-ray, ultrasonic and electronic apparatus if these are considered necessary. Detailed inspection and consultation at all stages would avoid undesirable modifications, e.g. introducing extra bends in pipe runs because of the prior installation of, say, electrical conduits.

OPERATION

The design should include recommendations for initial conditioning of the plant and subsequent operation and maintenance. This would cover details of the mechanical and corrosion limits of the materials used, methods for measuring the metal thickness in order to estimate corrosion losses and methods of dismantling the equipment for replacement of parts. Details should be given of corrosion stimulators which should be avoided, e.g. traces of mercury and copper salts with aluminium, zinc and iron or chromium salts with stainless steels as well as information on the corrosion inhibitors which may be used.

Other precautions to be pointed out include items such as the susceptibility of the equipment to local overheating. Details on the facilities provided and the procedure for draining the equipment during shut-down or alternatively filling with water should be given. Surface treatment, paints and the supervision of cathodic protection should be specified.

The frequency of periodic inspection is also important. It may often be desirable to shut down the entire plant every six months or so when all parts may be inspected and repaired, replaced or judged capable of lasting another six months. It is essential that the maximum accessibility of equipment should be assured in the design to make such inspection and maintenance easy.

Chapter XIII

DIAGNOSIS

In the diagnosis of corrosion troubles the first task is to assemble all
the necessary information regarding metal, environment and working
conditions. It is difficult for the non-corrosionist to appreciate all the
data essential for the solution of a particular problem. In this chapter all
the possible factors and the sequence of operations which have to be
followed in diagnosis are considered. This should also enable the user
to solve some of the more simple corrosion problems and to realise the
more important features.

At the outset it should be emphasised that the frequent tendency to blame
the quality of the metal is usually fallacious. Only in very rare cases
does this provide the answer to the problem.

GASES

In a few cases corrosion is accompanied by the evolution of gases which
may be detected by their odour or inflammability.

When the system is first opened up the gas most easily identified—by its
smell of rotten eggs—is sulphuretted hydrogen, i.e. hydrogen sulphide, H_2S.
This may be detected immediately or after disturbing corrosion products
and is usually indicative of the action of sulphate-reducing bacteria. The
smell does not persist indefinitely since the gas soon escapes and any sulphides, which evolve H_2S in acid, are oxidised to sulphate. Thus this piece
of evidence may easily be lost.

The use of cathodic protection in closed systems can also give rise to gas
evolution. If the current density is too high, chlorine can be generated at
the anode from sea water and waters high in chloride. This greenish-
yellow pungent gas is not only extremely corrosive but also toxic and
appropriate precautions should be taken.

In hot water heating systems, troubles are sometimes encountered due
to the accumulation of gas which interferes with water circulation. This
is often found to be inflammable which is usually a sure indication of the
presence of hydrogen.

Caution should be exercised in testing for any gases owing to the possibility of toxicity and explosion hazard.

SAMPLING

Samples of both the corroded metal and the water used are essential in
investigating any corrosion problem. When cutting a specimen, care must

be taken that the corroded surface is not damaged in any way. Thus, in the use of an oxy-acetylene torch the sample should be large enough to ensure that the heat does not affect the area to be examined. The sample cut out should be handled with great care so as not to damage the corrosion product adhering to the surface. It should be marked clearly so that the exact location and relationship to neighbouring plant is known. In the case of a section, the top and bottom should be clearly marked and also the direction of water flow. An initial examination should be made and details of the amount of corrosion product, its colour, form and smell recorded.

Every effort should be made to keep the specimen in its original form for subsequent detailed examination. A tube specimen can be filled with the corrosive medium and plugged at both ends while other specimens may be kept, fully immersed, in a suitable container.

A sample of the water and corrosion product should be taken in a clean glass or plastic bottle which should be well washed out with the water.

HISTORY OF FAILURE

In order to arrive at the factors contributing to the corrosion, a full history and description of the plant is usually required together with a rough line diagram. The latter should show the constructional materials used and the location of the corrosion-affected areas. Details of the flow velocities and temperatures in various parts of the system should be given. The age of the plant, any structural alterations and, above all, the condition of the original metals should be specified. In addition, the conditions of working should be fully described: the cycle of operations, the length of idle periods and the condition of the plant during these periods, any treatment by way of painting or cleaning, and, with cooling systems and boilers, the concentration factor and the amount and frequency of blow-down. Finally a full analysis of the water will be needed (as outlined in Chap. II) together with a record of any water treatment or pH control.

In describing the metal it is preferable if a standard specification can be given since trade names often hide an analysis which is difficult to trace. There is often some confusion of terms in describing metals, e.g. 'bronzes' are sometimes brasses and 'chromium steels' are variously referred to as stainless irons and stainless steels. It is always preferable if possible to give the composition and condition of the metal.

PRELIMINARY EXAMINATION

The first procedure is to examine carefully the corroded specimen, usually with the assistance of a low-powered lens. This examination, together with the information and case history supplied, is often adequate for a satisfactory diagnosis to be made. This examination is made in stages as the corrosion product is removed step by step until ultimately the metal surface is cleaned of corrosion product, often with the aid of dilute acid. Throughout this examination the amount and type of product and deposit on the surface and its colour are carefully noted.

Once the product has been removed it is then possible to deduce further information on the type and distribution of the attack.

CORROSION PRODUCTS AND DEPOSITS

If the product is very easy to remove it is suggestive that the corroding medium is a soft water. The absence of significant amounts of calcium carbonate may be shown on addition of a little acid when the gas evolution will be very small or negligible. If the product is difficult to remove even with strong acid it is likely to contain calcium sulphate or siliceous material. The occurrence of the latter is very noticeable with product on wrought iron. The thickness and nature of the deposit on the metal surface is affected not only by composition of water but also by temperature and speed of flow. For example, a heavy build-up of scale may be evidence either of a very hard water or a high temperature water.

If there is a considerable amount of product but only a small amount of attack on the underlying metal the corrosion has obviously taken place elsewhere in the system and the site of attack must be sought upstream. This type of product build-up may be caused by iron bacteria, which do not actively cause corrosion but accompany it by forming deposits which are more voluminous than normal corrosion product and may result in blockages.

COLOUR

Colours must be treated with some reserve as they are influenced so greatly by the mixing or state of division of products. The colour of the corrosion product on iron and steel is not usually of very great diagnostic value except in systems where the ingress of air is limited when it may give some indication of the state of oxidation. With increasing degree of oxidation the colours of the iron oxides and hydroxides range from the almost white ferrous hydroxide, $Fe(OH)_2$, through the 'green rusts', which vary from light to dark green, and black magnetite, Fe_3O_4, to the red-brown hydrated $Fe_2O_3 . H_2O$ or $FeOOH$ or anhydrous oxide, Fe_2O_3.

The corrosion products on copper and its alloys are usually green or blue and are usually indicative of the basic carbonate or sulphate. When impingement occurs on copper, the rounded and elongated pits which indicate the direction of flow are salmon red at first but, on exposure, develop a brown or purple film.

Although the colour of products on aluminium is usually white, a slight blue or greenish tinge may be associated with sites of pitting. This is usually indicative of the presence of copper as being an active cause of the localisation of attack. It should be emphasised, however, that the amount of copper necessary to initiate pitting on aluminium is very small and the absence of colour does not mean that this source of trouble can be eliminated.

SURFACE FILMS

On mechanical, or even acid, cleaning of the specimen an adherent film on the metal surface is sometimes revealed. This immediately suggests that the origin of the localised attack is a cathodic surface film. Two common ones are magnetite on iron and steel and cuprous oxide on copper. The magnetite on iron is easily recognised from its black or dark-grey colour as compared with the etched light-grey appearance of the rest of the surface. The thin film of cuprous oxide is readily recognised from its characteristic red colouration. The magnetite film may either be an original film of mill scale formed in hot-working of the steel or a more localised film produced in welding or other methods of fabrication in which heat has been applied. The red film on copper may likewise have been formed during manufacture or construction, by the use of an unsuitable priming paint such as red lead, or by heating during brazing.

The glassy red deposits sometimes seen on copper hot-water cylinders are cuprous oxide and can cause severe pitting in aggressive waters. Black films on copper may be cupric oxide or carbon and may be formed during fabrication.

In other cases localised attack may have been caused by an imperfect coating of a corrosion resistant metal on a less noble basis metal, e.g. tin, nickel or chromium on steel.

TYPE OF ATTACK

LOCALISATION

The type of attack may be general or localised. When the attack is general with even removal of metal over quite large areas producing grooving or thinning, this is usually indicative either of acid attack, e.g. attack on copper or steel in condensates or soft waters high in free carbon dioxide, or alkaline corrosion, e.g. with an amphoteric metal such as aluminium in alkaline water.

When the attack is more localised but still with quite large areas of wastage, it may be caused by bimetal couples, local turbulence or differential aeration. An example of this is where copper piping has been screwed into a galvanised tank and severe attack on the tank near the connection has occurred. In a more extreme case corrosion of an aluminium alloy tank had occurred near to where a brass baffle-plate had been bolted!

Severe corrosion in the bends of pipes or immediately downstream of screwed junctions and accompanied by large mounds of corrosion product is generally initiated by local turbulence.

Large patches of attack are often caused by differential aeration. This may occur in any crevice across the mouth of which aerated water is passing. A similar effect is produced by debris on stainless steel or aluminium bronze which need to remain clean to resist corrosion. Under marine conditions large corroded depressions on aluminium bronze can be caused by fouling.

PITTING

Pits may be essentially free from corrosion product or covered by a loose layer of easily removable product or covered by a hard nodule.

If the pits are essentially free from corrosion product their shape is instructive. Hemispherical pits are typical of cavitation, while pits which are elongated and undercut are typical of impingement attack under the action of local or general turbulence. Hemispherical pits, which are usually also free from product, may also occur owing to the influence of hot-wall or cold-wall effects, i.e. by the release of gas or steam bubbles on the metal wall.

Hemispherical pits beneath a covering of loose black product are often associated with the action of sulphate-reducing bacteria under anaerobic conditions. Confirmatory evidence is provided by the smell of hydrogen sulphide on opening the system or on acidifying the product. If necessary, a suspected case of this type of attack may be confirmed by bacteriological examination when the bacteria are grown in a suitable culture and identified under the microscope.

Pitting is almost invariably associated with the nodular form of product sometimes found on the interior of pipes. Such nodules are often very hard and difficult to remove when wet but on drying out they are brittle and often may be removed whole.

SELECTIVE ATTACK

Both graphitisation and dezincification may be recognised in the first place by the changes in colour which accompany them. Any yellow copper alloy such as brass which shows red patches has generally been dezincified by exposure to acidic or brackish water. Similar environments will cause the graphitisation of cast iron as shown by a black film or layer on the grey iron. The affected part can be cut easily with a knife and will mark paper in a similar way to a pencil.

FURTHER EXAMINATION

In a number of cases there may still be some doubt as to the cause of the particular type of attack and, even when the diagnosis seems reasonable, confirmatory tests may be required. These may involve further examination of the corrosion product or microscopic examination of polished metal sections.

The chemical tests can be very simple as is exemplified by the case of
a galvanised cooler which had undergone heavy nodular attack when there
was reason to suspect the presence of copper. The nodules were carefully
prised free from the surface and their cores, i.e. the part of the product
near the bottom of the pit, separated. Electrolysis of an acid solution of
these between two platinum wires revealed the presence of an appreciable
amount of copper which was deposited on the surface of the cathode.
This example also serves to emphasise the care needed in the examination and the separation of different parts of the corrosion product.

In a second example a minor explosion had occurred in a water softener,
presumably due to the accumulation of an inflammable gas. This was
found to be due to a residue of dross from the galvanising bath and
enough of this residue remained to prove that it would generate hydrogen
freely even in distilled water.

X-ray examination may be useful in certain cases. Often, when there is
still some doubt about the cause of corrosion, it is necessary to take a
powder photograph of the product or deposit. This may yield the required
information on the composition of the product and indicate, for example,
the temperature at which the corrosion occurred. This method of examination may also be valuable where there has been some carry-over from
one part of the system to another, as from a boiler to a condenser.

In metallurgical examinations it is desirable to retain the corrosion product on the metal surface. This is best effected by immersing the specimen in a cold-setting resin contained in a mould which is then placed in
a desiccator and evacuated. Air is extracted and release of the vacuum
allows the resin to penetrate into the product and fix it in position when
set. Both longitudinal and transverse sections should be subsequently
polished. Such examinations are used to confirm selective corrosion or
to indicate the combined influence of mechanical and corrosive factors
as in stress corrosion cracking or corrosion fatigue. In the examination
of cracking and weld failures this procedure would almost automatically
be carried out.

BIBLIOGRAPHY

GENERAL

Evans, U. R. (1960). *The Corrosion and Oxidation of Metals*. Arnold, London. pp. 1094.

Shreir, L. L. (Ed.) (1963). *Corrosion*. Newnes, London. pp. 1738.

Uhlig, H. H. (Ed.) (1948). *The Corrosion Handbook*. Wiley, New York. Chapman and Hall, London. pp. 1188.

LaQue, F. L. and Copson, H. R. (Eds.) (1963). *Corrosion Resistance of Metals and Alloys*. 2nd. Edt. Reinhold, New York. pp. 712.

Tödt, F. (Ed.) (1961). *Korrosion und Korrosionschutz*. de Gruyter, Berlin. pp. 1427.

Tomashov, N. D. (1959). *Theory of Corrosion and Protection of Metals*. Academy of Sciences, USSR, Moscow. pp. 592.

Rosenfeld, I. L. (1960). *Atmospheric Corrosion of Metals*. Academy of Sciences, USSR. Moscow. pp. 372.

Klas, H. and Steinrath, H. (1956). *Die Korrosion des Eisens und ihre Verhütung*. Verlag Stahleisen, M. B. H., Düsseldorf.

Champion, F. A. (1965). *Corrosion Testing Procedures*. 2nd Edt. Chapman and Hall, London.

Annual Reports on the Progress of Applied Chemistry. Corrosion Section. Soc. of Chem. Ind. London.

Wormwell, F. and Evans, E. Ll. (1956). *Corrosion of Metals. Chemical Engineering Practice* (Eds. Cremer, H. W. and Davies, T.) **2**. Chap. 8. pp. 255-341. Butterworths (12 vols.)

INTRODUCTION

Vernon, W. H. J. (1949). *The Corrosion of Metals*. J. Roy. Soc. Arts. **97**. pp. 578-610.

Vernon, W. H. J. (1956-7). *Metallic Corrosion and Conservation. Inst. Civil Engrs*. pp. 105-133.

Uhlig, H. H. (1950). *Proceedings of the United Nations Scientific Conference on the Conservation and Utilisation of Reserves*. **2**. p. 213.

Warner, H. K. (1956). The Cost of Corrosion. *Corrosion, A Symposium*. University of Melbourne. pp. 1-14.

Uhlig. H. H. (1950). The Cost of Corrosion to the United States. *Corrosion*. 6. pp. 29-33. *Chemical and Engineering News*. (1949). **27**. pp. 2764-67.

Wachter, A. (1954). Relation of Corrosion to Business Costs. *Corrosion*. **10**. pp. 273-8.

Bhagavantam, S. (1964). *Commonwealth Symposium*. New Delhi, India.

CHAPTER I

Turnbull, A. H., and Davis, H. C. (1942). Electrode Potentials of Metals in Sea Water. *Aeronaut Res. Council Technical Report*. R and M 1901. HMSO, London.

Elze, J. and Oelsner, G. (1959). The Electrochemical Series of Metals in Practical Corroding Media. *Metalloberfläche*. 5. 129.

Chilton, J. P. (1961). *Principles of Metallic Corrosion*. Roy. Inst. Chem., London. pp. 64.

Evans, U. R. (1963). *An Introduction to Metallic Corrosion*. 2nd Edt. Arnold, London. pp. 253.

CHAPTER II

Hamer, P., Jackson, J. and Thurston, E. F. (1961). *Industrial Water Treatment Practice*. Butterworths and I.C.I. Ltd. London. pp. 514.

Nordell, E. (1961). *Water Treatment*. 2nd. Edt. Reinhold, New York. Chapman and Hall, London. pp. 526.

Powell, S. T. (1954). *Water Conditioning for Industry*. McGraw-Hill, New York and London.

British Waterworks Year Book and Directory (1961). Part 3. Millis, L. (Ed.). pp. 500.

CHAPTERS IV and V

Rabald, E. (1951). *Corrosion Guide*. Elsevier, New York. pp. 629.

Ritter, F. (1952). *Korrosionstabellen Metallischer Werkstoffe*. Springer-Verlag, Vienna. pp. 283.

Shell Development Co. (1960). *Corrosion Data Survey*. Emeryville, California. U.S.A. pp. 60 and 97 charts.

Kenworthy, L. (1954). Corrosion of Machinery in H.M. Ships. *J. Appl. Chem.* 102.

Kenworthy, L. (1965). Some Corrosion Problems in Naval Marine Engineering. *Inst. Mar. Engrs.* 77. 149

Vernon, W. H. J. (1934). Basic Copper Carbonate and Green Patina. *J. Chem. Soc.* Nov. p. 1853

Inst. Civil Engrs. (1940). *Deterioration of Structures in Sea Water.* 18th. Report. London. pp. 52.

CHAPTER VI

Evans. U. R. and Rance, V. (1958). *Corrosion and its Prevention at Bimetallic Contacts.* HMSO, London.

CHAPTER VII

Robertson, W. D. (1956). *Stress Corrosion Cracking and Embrittlement.* Wiley, New York. Chapman and Hall, London. pp. 202.

Fearon, W. (1964). Waterside Attack of Diesel Engine Cylinder Liners. *J. Roy. Nav. Scient. Serv.* **19.** 1. Jan.

Godfrey, D. J. (1959). Cavitation Damage: A Review of Present Knowledge. *Chem and Ind.* June 6th. p. 686

Godfrey, D. J. (1964). Studying Cavitation Damage of Materials. *J. Roy. Nav. Scient. Serv.* **19.** 3.

Peiser, H. S. and Tytell, B. H. (1961). The Electrochemical Approach to Cavitation Damage and its Prevention. *Corrosion.* **17.** 535t.

Collins, H. H. (1963). Cathodic Protection of Cast Iron Propellers. *Brit. Cast Iron Res. Assoc. Journal.* **11.** No. 5. p. 623.

N.P.L., Teddington. (1956). *Cavitation in Hydrodynamics.* Proc. Symposium. Sept, 1955. HMSO, London. pp. 458.

CHAPTER VIII

McAdams, W. H. (1954). *Heat Transmission.* 3rd Edt. McGraw-Hill, New York.

CHAPTER IX

Bregman, J. I. (1963). *Corrosion Inhibitors.* MacMillan, New York. Collier-MacMillan, London. pp. 320

Putilova, I. N., Balezin, S. A. and Barannik, V. P. (1960). *Metallic Corrosion Inhibitors.* Pergamon, London. pp. 191.

CHAPTER X

Cartwright, P. A., (1950). *Metal Finishing Handbook.* Blackie, London and Glasgow. pp. 216.

Ballard, W. E. (1948). *Metal Spraying and Sprayed Metal.* 3rd. Edt. Griffin, London. pp. 362.

Gailer, J. W. and Vaughan, E. J. (1950). *Protective Coatings For Metals.* Griffin, London. pp. 261.

Fedot'ev, N. P. and Griklikhes, S. Ya. (1959). *Electropolishing, Anodising and Electrolytic Pickling of Metals.* Translation by Draper, Teddington. London. pp. 285.

Fancutt, F. and Hudson, J. C. (1957). *Protective Painting of Structural Steel.* Chapman and Hall, London. pp. 102.

Rischbieth, J. R. and Marson, F. (1962). Ship Bottom Paints Based on Sodium Silicate. *Fourth Australian O. C. C. A. Convention.*

Preston, R. St. J. (1946). Bitumen and Bituminous Coatings. *Chem. Res. Spec. Report.* HMSO, London.

Mayne, J. E. O. and Thornhill, R. S. (1948). Cementiferous Paints. *J. Iron Steel Inst.* **158.** 219.

Pyefinch, K. A. (1948). Marine Exposure of Cementiferous Painting Schemes. *J. Iron Steel Inst.* **158.** 229.

CHAPTER XI

Morgan, J. H. (1959). *Cathodic Protection.* Leonard Hill, London, pp. 325.

Applegate, L. M. (1960). *Cathodic Protection.* McGraw-Hill, New York. pp. 229.

Symposium (1964). *Recent Advances in Cathodic Protection.* Marston Excelsior Ltd., Birmingham. May.

Symposium (1949). *Cathodic Protection.* N.A.C.E., Houston, U.S.A. pp. 203.

Symposium (1953). *Cathodic Protection.* Corrosion Group, Soc. Chem. Ind. London. pp. 65.

Rohrman, F. A. (1950). *Bibliography of Cathodic Protection. World Oil.* **131.** 179-180. 436 refs.

Crennell, J. T., and Wheeler, W. C. G. (1956). Zinc Anodes For Use in Sea Water. *J. Appl. Chem.* **Oct.** 415.

Baptista, F. G. and Finley, H. F. (1963). World Petroleum Congress. Corrosion Problems in Lake Maracaibo. *Corr. Prev. and Control.* **10.** No. 9. p. 31.

Edeleanu, C. (1954). Anodic Protection. *Metallurgia.* **50.** 113.

Edeleanu, C. (1961). Anodic Protection. *Chem and Ind.* **Mar.** 301.

CHAPTER XIII

Cotton, J. B. (1956). Examination of Corroded Specimens. *Corr. Tech.* **3.** No. 5. May.

Cotton, J. B. and Watkins, R. J. (1957). Examination of Corroded Specimens. *Corr. Tech.* **4.** No. 2. Feb.

APPENDIX

CONVERSION FACTORS

Length

Multiply	*By*	*To Obtain*
Angstrom units	3.937×10^{-9}	inches
Angstrom units	1.0×10^{-10}	metres
Angstrom units	1.0×10^{-4}	microns
centimetres	1.0×10^{8}	Angstrom units
centimetres	0.3937	inches
feet	0.3048	metres
inches	2.54	centimetres
metres	3.281	feet
microns	1.0×10^{-4}	centimetres
microns	0.03937	mils
millimetres	39.37	mils
mils	0.00254	centimetres

Area

square centimetres	0.155	square inches
square feet	0.093	square metres
square inches	6.45	square centimetres
square metres	10.76	square feet

Volume

cubic centimetres	0.061	cubic inches
cubic feet	0.0283	cubic metres
cubic feet	6.23	gallons
cubic feet	28.32	litres
cubic inches	16.39	cubic centimetres
cubic metres	35.315	cubic feet
Imperial gallons	4.546	litres
Imperial gallons	0.1605	cubic feet
litres	0.0353	cubic feet
litres	0.22	gallons
litres	1.76	pints
pints	0.568	litres

Weight

grams	0.002205	pounds (av.)
grams	0.035274	ounces (av.)

Multiply	By	To Obtain
kilograms	2.2046	pounds (av.)
ounces (av.)	437.5	grains
ounces (av.)	28.35	grams
pounds (av.)	0.4536	kilograms
pounds per cubic foot	0.016	grams per c.c.

Pressure

atmospheres	1.033	kilograms per sq. cm.
atmospheres	14.696	lb per sq. in.
mm of mercury	0.001316	atmospheres
inches, mercury gauge	0.0334	atmospheres
inches, mercury gauge	13.59	inches, water gauge
inches, mercury gauge	0.4912	lb per sq. in.
inches, water gauge	0.07355	inches, mercury gauge
inches, water gauge	0.036	lb per sq. in.
kilograms per sq. cm.	14.223	lb per sq. in.
kilograms per sq. metre	0.2048	lb per sq. ft.
lb per sq. ft.	4.8824	kilos per sq. metre
lb per sq. in.	0.0703	kilos per sq. cm.

Velocity

feet per second	60.0	feet per minute
feet per minute	0.01667	feet per second
feet per second	0.6818	miles per hour
miles per hour	1.4667	feet per second
centimetre per second	0.03281	feet per second

Electrical resistance

ohms per cm. cube	0.3937	ohms per in. cube
ohms per in. cube	2.54	ohms per cm. cube

WATER

Concentration

grains per gallon	1.4254	parts per 100,000
grains per gallon	14.254	parts per million, ppm.
equivalents per million	0.001 × sp. gr.	normality
equivalents per million	50.0	parts per million, $CaCO_3$
grams per litre	1,000 ÷ sp. gr.	ppm
mls. dissolved oxygen per litre (NTP)	1.429 ÷ sp. gr.	ppm
ppm	0.001 × sp. gr.	grams per litre
ppm dissolved oxygen	0.70 × sp. gr.	mls. dissolved oxygen per litre (NTP)
ppm	0.07	grains per gallon
parts per 100,000	0.7	grains per gallon

Hardness

Degrees hardness:-
    English degrees (Clark) = grains calcium carbonate per gallon or
parts per 70,000.
    French degrees (Continental degrees) = parts of calcium carbonate
per 100,000.
    German degrees = parts of calcium oxide per 100,000.

Multiply	*By*	*To Obtain*
English degrees	1.43	French degrees
English degrees	0.8	German degrees
French degrees	0.7	English degrees
German degrees	1.24	English degrees

Hydraulic equivalents

One Imperial gallon	=	277.41 cubic ins.
	=	0.1605 cubic foot
	=	10.0 lb
	=	4.546 litres
One cubic inch of water	=	0.03608 lb
	=	0.003605 gal.
One cubic foot of water	=	6.23 gallons
	=	28.317 litres.
	=	0.0283 cubic metre
	=	62.30 lb
	=	0.557 cwt.
	=	0.028 ton.
One pound of water	=	27.68 cubic ins.
	=	0.10 gallon
	=	0.4537 kilo
One cwt. of water	=	11.2 gallons
	=	1.8 cubic feet
One ton of water	=	35.9 cubic feet
	=	224.0 gallons
	=	1,000.0 litres (approx)
	=	1.0 cubic metre (approx)
One litre of water	=	0.22 gallon
	=	61.0 cubic inches
	=	0.0353 cubic foot
One cubic metre of water	=	220.0 gallons.
	=	1.308 cubic yards
	=	61,028.0 cubic inches
	=	35.31 cubic feet
	=	1,000.0 kilos
	=	1.0 ton (approx)
	=	1,000.0 litres

One kilo of water	=	2.204 lb
One atmosphere	=	1.054 kilo per sq. in.
A column of water 1 ft. high	=	Pressure of 0.434 lb per sq. in.

Contents in Gallons of Various Sizes of Pipes each One Foot Long

$\frac{1}{2}$ in. pipe	0.0084 gallons
1 " "	0.0339 "
$1\frac{1}{4}$ " "	0.0530 "
$1\frac{1}{2}$ " "	0.0763 "
2 " "	0.1356 "
$2\frac{1}{2}$ " "	0.2120 "
3 " "	0.3053 "
4 " "	0.5426 "
5 " "	0.848 "
6 " "	1.221 "
7 " "	1.662 "
8 " "	2.171 "
9 " "	2.747 "
10 " "	3.393 "
11 " "	4.105 "
12 " "	4.881 "

CORROSION UNITS, CONVERSION FACTORS

Multiply	*By*	*To Obtain*
Ins. per year (ipy)	696 × density	mg. per sq. dm. per day (mdd)
mdd	0.00144 ÷ density	ipy
gm. per sq. metre per day	0.0144 ÷ density	ipy

Approximate Factors for Converting mdd to mpy

Magnesium	0.83
Aluminium alloys	0.53
Zinc	0.20
Iron alloys	0.19
Nickel alloys	0.16
Copper alloys	0.16
Lead	0.13

Useful Conversion Factors to Remember.

One cu. ft.	=	6.23 gallons
One gallon	=	10.0 lb
One lb.	=	454.0 grams
60 mph	=	88.0 ft. per sec.
One U.S. gallon	=	8.33 lb
One cubic foot	=	7.48 U.S. gallons